NEWCOMERSTOWN PUBLIC LIB
123 N. BRIDGE STREET
NEWCOMERSTOWN, OH 43832

DISCARD

j520
Asi

91-1326

Asimov, Isaac
    Isaac Asimov's library
of the universe
              Index

NEWCOMERSTOWN PUBLIC LIBRARY
123 N. Bridge St.
Newcomerstown, Ohio 43832-1093

ISAAC ASIMOV'S
Library of the Universe

Index

Gareth Stevens Publishing
Milwaukee

Cover background photography: National Optical Astronomy Observatories. The reproduction rights to all other photographs and illustrations on the cover of this book are controlled by and may not be reproduced without the permission of the individuals or institutions credited on page 32 of each of the respective titles in this series.

Library of Congress Cataloging-in-Publication Data

Asimov, Isaac, 1920-
    Isaac Asimov's library of the universe, index / by Isaac Asimov.
        p.   cm.
    Summary: A comprehensive index to the complete set of Isaac Asimov's thirty-two volume Library of the Universe, a series of space science books that introduce young readers to the facts and mysteries of the cosmos.
    ISBN 1-55532-900-4
    1. Asimov, Isaac, 1920-   Library of the universe—Indexes—Juvenile literature.
    [1. Asimov, Isaac, 1920-   Library of the universe—Indexes.]   I. Title.
QB981.A933A85   1990
520—dc20   89-43142

A Gareth Stevens Children's Books edition

Edited, designed, and produced by
Gareth Stevens, Inc.
RiverCenter Building, Suite 201
1555 North RiverCenter Drive
Milwaukee, Wisconsin   53212, USA

Index copyright © 1990 by Gareth Stevens, Inc.
Format copyright © 1990 by Gareth Stevens, Inc.

First published in the United States and Canada in 1990 by Gareth Stevens, Inc. All rights reserved. No part of this book may be reproduced or used in any form or by any means without permission in writing from Gareth Stevens, Inc.

For a free color catalog describing Gareth Stevens' list of high-quality children's books, call 1-800-341-3569 (USA) or 1-800-461-9120 (Canada).

Cover design: Kate Kriege, © Gareth Stevens, Inc., 1990

Index editor: John D. Rateliff
Project editor: Mark Sachner
Research editor: Scott Enk
Series design: Laurie Shock
Book design: Kate Kriege

Printed in the United States of America

1 2 3 4 5 6 7 8 9 96 95 94 93 92 91 90

A note about using this index

The purpose of this *Index* volume is to give you a quick and complete guide to the other 32 volumes of Isaac Asimov's Library of the Universe.

Each entry in this index shows you where to look to find out about the many topics and ideas contained in the Library of the Universe. This index also offers you a big plus: It gives brief explanations of what many of the entries mean.

For example, "Apollo" is the name of an ancient Greek god, an asteroid, and a series of Moon flights. So after each entry for "Apollo," you can find out whether the entry refers to the god, the asteroid, or something else. Here is an example:

>Apollo (Greek god)
>Apollo (asteroid)

There are other references using the word "Apollo," such as Apollo 11 (the first Moon landing) and the Apollo-Soyuz Test Project (a joint US-Soviet flight). So in addition to showing you where to go to learn more about Apollo, the index teaches you something about the many meanings — and uses — of the word.

Before using this index, look at the list of book titles below. Opposite the title of each book is a code form of that title:

TITLE	CODE	TITLE	CODE
Ancient Astronomy	ANCIENT	*Our Milky Way and Other Galaxies*	GAL
The Asteroids	ASTER	*Our Solar System*	SOL
Astronomy Today	TODAY	*Piloted Space Flights*	PILOT
The Birth and Death of Stars	STAR	*Pluto: A Double Planet?*	PLUTO
Colonizing the Planets and Stars	COL	*Projects in Astronomy*	PROJ
Comets and Meteors	COMET	*Quasars, Pulsars, and Black Holes*	QPB
Did Comets Kill the Dinosaurs?	DINOS	*Rockets, Probes, and Satellites*	RPS
Earth: Our Home Base	EARTH	*Saturn: The Ringed Beauty*	SAT
The Earth's Moon	MOON	*Science Fiction, Science Fact*	SF
How Was the Universe Born?	UNIV	*Space Garbage*	GARB
Is There Life on Other Planets?	LIFE	*The Space Spotter's Guide*	GUIDE
Jupiter: The Spotted Giant	JUP	*The Sun*	SUN
Mars: Our Mysterious Neighbor	MARS	*Unidentified Flying Objects*	UFO
Mercury: The Quick Planet	MERC	*Uranus: The Sideways Planet*	UR
Mythology and the Universe	MYTH	*Venus: A Shrouded Mystery*	VENUS
Neptune: The Farthest Giant	NEP	*The World's Space Programs*	WSP

Throughout the index, you can first identify the books in which a word, term, or idea appears, and then the pages on which it appears. Here is a sample:

>Apollo missions COL 15; GARB 6-7; MOON 7, 13, 14, 16-17, 19, 22; PILOT 9, 10-11, 16-17, 18; RPS 13; SF 12-13; WSP 6-7, 11

In this entry, you find Apollo missions mentioned in seven books: in *Colonizing the Planets and Stars* (on page 15), in *Space Garbage* (on pages 6-7), in *The Earth's Moon* (on pages 7, 13, 14, 16-17, 19, and 22), in *Piloted Space Flights* (on pages 9, 10-11, 16-17, and 18), in *Rockets, Probes, and Satellites* (on page 13), in *Science Fiction, Science Fact* (on pages 12-13), and in *The World's Space Programs* (on pages 6-7 and 11).

A

A-1 (*see* Asterix)
Accretion disks GAL 8, 25; QPB 10-11, 18-19, 28-29
Acid rain VENUS 17
Adams, John Couch (British astronomer) NEP 5, 25
Adrastea (moon of Jupiter) JUP 14-15, 28
Afghanistan WSP 29
Africa COL 5; EARTH 8; LIFE 15; MOON 13; RPS 4; VENUS 21
Age of Reptiles DINOS 28-29; VENUS 10
Akhenaton (Egyptian pharaoh) MYTH 5
Al-Battani (Arabic astronomer) ANCIENT 20
Al-Saud, Prince Sultan Abdul Aziz WSP 29
Alaska SUN 18, 19
Alberta ASTER 18
Aldebaran (star) GUIDE 16-17; STAR 13
Alderamin (former pole star) MYTH 25
Aldrin, Edwin "Buzz" (astronaut) MOON 16; PILOT 10-11; SF 13; WSP 7 (*see also* Apollo missions: Apollo 11)
Algol (star) MYTH 25
Alien (movie) PILOT 24-25
Alien life ASTER 27; COL 27; GARB 27; PILOT 11; PROJ 4; QPB 13, 14; SF 6-7, 19, 22-23; TODAY 11; UFO 17, 18-19, 20-21, 24-25, 26-27, 29
  On Mars LIFE 10, 12-13; MARS 4-11, 12, 14, 25; RPS 22-23; UFO 17
  On the Moon LIFE 10-11; UFO 24
  On Venus UFO 29
  (*see also the book* Is There Life on Other Planets?)
Allen, Joe (astronaut) RPS 10, 11
Alligators DINOS 5
Almanacs PROJ 10, 16, 20
Alouette 1 (satellite) WSP 28
Alpha Centauri (star) COL 18, 28; GUIDE 23; LIFE 17; MYTH 17; NEP 27; SOL 11, 21, 27; SUN 28; UNIV 14, 22-23, 29
Altair (star) GUIDE 12-13
Amalthea (moon of Jupiter) JUP 15, 28-29; TODAY 23
Amazing Stories (science fiction magazine) SF 6-7, 28-29 (*see also* Gernsback, Hugo)
Amazon Basin (Brazil) RPS 13
American Indians (*see* Indians, American, *and the names of individual tribes and groups*)
Americas RPS 4 (*see also* North America *and* South America)
Ammonia JUP 8-9; LIFE 5, 14, 17, 18; SAT 25; SOL 14

Ananke (moon of Jupiter) JUP 14-15, 29
Anasazi Indians ANCIENT 8
Andromeda (mythological figure) MYTH 28
Andromeda (constellation) GUIDE 14-15; MYTH 28
Andromeda Galaxy COL 19; GAL 14-15, 20, 26; GUIDE 14-15, 26-27; LIFE 26; PROJ 20; QPB 17, 27, 29; TODAY 10-11, 15; UNIV 8-9, 14-15, 22-23, 29
ANS-1 (*see* Astronomical Netherlands Satellite)
Antarctic Circle COMET 13; MYTH 5
Antarctica COMET 7, 13; EARTH 23; JUP 20-21; LIFE 17, 22-23; MARS 12; MOON 22; RPS 12
Antares (star) GUIDE 12-13
"Antennae" galaxies GAL 26
Anti-gravity device WSP 25
Antlia (constellation) GUIDE 29
Apache Point Telescope (New Mexico) TODAY 17
Aphrodite (Greek goddess of love) VENUS 23
Aphrodite Terra ("continent" on Venus) VENUS 21, 22-23
Apollo (Greek god) MYTH 5
Apollo (asteroid) ASTER 18
Apollo missions COL 15; GARB 6-7; MOON 7, 13, 14, 16-17, 19, 22; PILOT 9, 10-11, 16-17, 18; RPS 13; SF 12-13; WSP 6-7, 11
  Apollo 7 RPS 13
  Apollo 8 MOON 7
  Apollo 9 MOON 16; RPS 13
  Apollo 11 MOON 13, 16, 17, 19; PILOT 10-11; SF 13; UFO 15
  Apollo 12 MOON 17; PILOT 9
  Apollo 13 MOON 14
  Apollo 14 WSP 11
  Apollo 15 MOON 17, 19
  Apollo 16 MOON 19
  Apollo 17 MOON 13
Apollo-Soyuz Test Project PILOT 16-17
Apple MERC 27
Apus (constellation) GUIDE 29
Aquarius (constellation) GUIDE 19
Aquila (constellation) GUIDE 12-13
Arabs, ancient ANCIENT 11, 20-21, 28; MYTH 25; QPB 12 (*see also* Muslims)
Arabsat (satellites) WSP 29
Aranca (former constellation) MYTH 28-29
Arc rings (*see* Neptune: Rings of)
Arctic Circle COMET 13; LIFE 28; MYTH 5
Arcturus (star) GUIDE 10-11
Arecibo Message SF 22-23
Arecibo radio telescope (Puerto Rico) SF 22-23
Arethusa (asteroid) ASTER 28

Argon (gas) MARS 12
Ariel (fairy) UR 10
Ariel (moon of Uranus) UR 10-11, 21, 22, 28-29
Ariel 1 (satellite) WSP 28
Aries (constellation) GUIDE 19
Arizona (US) ASTER 18; COMET 4-5, 25; DINOS 8; GAL 4-5; MARS 11; QPB 25; RPS 15; SOL 20; SUN 18
Armstrong, Neil (astronaut) MOON 16; PILOT 11; RPS 16-17; WSP 5 (see also Apollo missions: Apollo 11)
Arnold, Kenneth (UFO spotter) UFO 5, 23
Artemis (Greek goddess of the Moon) MYTH 6
Artificial atmospheres (see Atmospheres, artificial)
Artificial gravity (see G-force and Gravity, artificial)
Aryabhata (satellite) WSP 29
Ashen Light VENUS 25
Asia EARTH 8, 10; RPS 4 (see also Australasia)
Asimov, Isaac SF 7
Asterio Regio (region of Venus) VENUS 10
Asterix (satellite) WSP 28
Asteroid belt ASTER 4, 6-7, 8-9, 12, 14; COL 11, 17, 23, 28-29; DINOS 12; LIFE 16; MARS 28; NEP 7; SOL 21
"Asteroid people" ASTER 24, 26-27; COL 10-11, 17, 22-23; WSP 19 (see also Space colonies)
Asteroids COL 10-11, 17, 22-23, 28-29; COMET 12, 25; DINOS 12-13, 14, 16, 18, 24-25, 29; GARB 28-29; GUIDE 26; JUP 9, 15; LIFE 16; MARS 25, 27, 28; MERC 25; MOON 20-21; MYTH 11; NEP 20, 24-25; PLUTO 15, 27; PROJ 4, 12; QPB 19; RPS 19, 26; SAT 19, 23; SOL 20-21; SUN 8; WSP 19
  As a broken planet ASTER 25
  Discovery of ASTER 6-7, 9, 11, 15, 28; MYTH 11
  Lost ASTER 28; GUIDE 26
  As moons (see Moons: Captured asteroids as)
  Outside of asteroid belt ASTER 14, 16, 18-19; DINOS 12
  (see also the names of individual asteroids and the book **The Asteroids**)
"Astrograms" TODAY 19
Astrolabe ANCIENT 21, 28
Astrology ANCIENT 13; MYTH 9, 26-27
Astronaut training PILOT 18-19
Astronauts and cosmonauts COL 7, 10-11, 12-13, 15, 17, 24-25; EARTH 18, 29; GARB 6-7, 8-9, 12, 19; MARS 20, 22-23, 24-25; MERC 16-17; MOON 16-17; PROJ 4; RPS 10-11, 16-17, 27; SF 10-11, 12-13, 24; SUN 9, 24; WSP 5, 6, 7, 13, 15, 16-17, 23, 25, 28-29 (see also names of individual astronauts and cosmonauts and the book **Piloted Space Flights**)
Astronomical Netherlands Satellite (satellite) WSP 29
Atlantic Ocean EARTH 9; PILOT 4-5
Atlantis (space shuttle) VENUS 23; WSP 29
Atlas (mythological figure) MYTH 11
Atlas (moon of Saturn) MYTH 11; SAT 16, 28-29
Atmospheres (see subentries under the names of individual planets and moons)
Atmospheres, artificial MARS 23; MOON 27; PILOT 20
Atmospheric pressure VENUS 15, 17
Atomic bomb and atomic energy QPB 11; SF 4, 7, 28
Aton (Egyptian Sun god) MYTH 5
Aurora EARTH 14, 16-17; JUP 24-25; SUN 18-19; VENUS 25 (see also Aurora Australis and Aurora Borealis)
Aurora Australia (Southern Lights) SUN 19
Aurora Borealis (Northern Lights) SUN 19
Australasia RPS 4 (see also Australia, Indonesia, and New Zealand)
Australia COL 26; DINOS 9; GARB 16; TODAY 29; WSP 28, 29 (see also Australasia)
Axial tilt (see subentries under Earth, Mars, Neptune, Saturn, Uranus, and Venus)
Azores (islands) EARTH 8
Aztecs ANCIENT 9; COMET 20; MYTH 19
Azur (satellite) WSP 29

B

Babel (see Tower of Babel)
Babylonians ANCIENT 12-13, 15; COMET 28; MYTH 5, 8, 17, 19, 23, 28; VENUS 7, 23
Bacon, Sir Francis (British author) SF 28
Bacteria COL 29; LIFE 4, 22-23, 28
Badlands, the (South Dakota) PROJ 14
Ball lightning UFO 14
Balloons WSP 4, 25
Barnard, E. E. (US astronomer) TODAY 23
Barringer Meteor Crater (Arizona) ASTER 18; COMET 25; DINOS 8
Basel (Switzerland) UFO 6
Baum, Richard M. (British astronomer) VENUS 19
Bayer, Johann (German astronomer) GUIDE 29
Bean, Alan (astronaut) MOON 17
Bell, Jocelyn (British astronomer) QPB 13; TODAY 11, 29

Bellamy, Edward (US author) SF 28 (see also *Looking Backward, 2000-1887*)
Belleville, Wisconsin (US) UFO 11
Belts (see subentries under Jupiter and Saturn)
Bergerac, Cyrano de (French author) SF 8, 24-25
Beta Centauri (star) STAR 12
Beta Pictoris (star) LIFE 17, 24; SOL 26; TODAY 10-11; UNIV 24
Betelgeuse (star) GUIDE 16-17, 25; STAR 12, 13, 14; SUN 21
Bethlehem (see Star of Bethlehem)
Bible ANCIENT 12-13; UFO 6-7, 25; VENUS 7 (see also Elijah, Ezekiel, Isaiah, Star of Bethlehem, and Tower of Babel)
Big Bang STAR 4-5, 6, 27; TODAY 29; UNIV 18, 20-22, 24, 26-27
"Big Crunch" UNIV 26-27
Big Dipper (constellation) ANCIENT 6; GUIDE 8-9, 10-11; MYTH 21, 23, 24-25, 29; PROJ 20
Big Horn (Wyoming) TODAY 5
Binary stars GAL 10, 25; STAR 21 (see also Double-star system[s] and Multiple stars)
Binoculars GUIDE 24; JUP 5, 6-7; MOON 10; PROJ 20-21, 26; TODAY 21
Biosphere II (environmental experiment) COL 6
Birds COL 9; DINOS 5, 25; LIFE 22; UFO 29; WSP 19
  As possible dinosaurs DINOS 25
Black dwarfs (stars) SOL 29; STAR 24, 29
Black holes GAL 9, 13, 20, 24-25; GARB 23; LIFE 17; QPB 8-9, 15, 16-21, 22, 26, 28-29; RPS 29; STAR 20; TODAY 11, 28, 29; UNIV 3, 26-27; WSP 23
Black Stone of the Kaaba MYTH 13
  As a possible meteorite MYTH 13
Blanc, Mont MARS 13
Blue giants (stars) GAL 18
Bode, Johann (German astronomer) UR 6
Bode's Law ASTER 6; UR 6
Bonestell, Chesley (US artist) SF 17
Book of Ezekiel (see Ezekiel)
Boötes (constellation) GUIDE 10-11; PROJ 16
Bradfield, William (comet spotter) TODAY 19
Brahe, Tycho (Danish astronomer) ANCIENT 24-25, 28-29
Brasilsat 1 (satellite) WSP 29
Braun, Wernher von (see Von Braun, Wernher)
Brave New World (novel) SF 29 (see also Huxley, Aldous)
Brazil RPS 13; UFO 19; WSP 29
Britain and the British COMET 21, 28; GUIDE 6; MOON 7; MYTH 11; NEP 5, 13; PLUTO 6; PROJ 24; QPB 22; RPS 28-29; SF 12; TODAY 11, 29; UFO 8, 12-13, 22; UR 5, 10, 11; WSP 14, 15, 28
Broken planet(s)
  Asteroids as ASTER 25
  Earth-Moon as MOON 20-21
  Pluto-Charon as PLUTO 15
Brown dwarfs (stars) STAR 29
Brucia (asteroid) ASTER 28
Bufo (former constellation) MYTH 28-29
Bug Nebula STAR 25
Building blocks of life LIFE 4-5, 9, 16-17
Bulgaria WSP 29
Bunsen, Robert Wilhelm (German scientist) TODAY 29
Buran (Soviet space shuttle) PILOT 13
Burnell, Jocelyn Bell (see Bell, Jocelyn)
Burney, Venetia (British schoolgirl) MYTH 11; PLUTO 6
Bussard Ramjet Starship (proposed spacecraft) COL 20

C

Calendars ANCIENT 4, 7, 11; MOON 8, 20; MYTH 7, 19; PROJ 10, 24
California (US) EARTH 11; MYTH 29; VENUS 22
Callisto (Galilean moon of Jupiter) JUP 14-15, 16-17, 18, 28-29; RPS 25; SOL 13, 19
Caloris basin (crater on Mercury) MERC 12-13, 15, 27
Caloris impact, the MERC 13
Calypso (moon of Saturn) SAT 22, 28-29
Cambridge, Massachusetts (US) TODAY 19
Canada EARTH 21; GARB 16-17; RPS 11, 28-29; SF 24; SUN 18; WSP 9, 11, 28
Canals, "Martian" LIFE 10, 14; MARS 6-7, 9, 11; RPS 22-23; SF 14
Cancer (constellation) GUIDE 19
Canis Major (constellation) GUIDE 16-17; MYTH 5
"Cannibal" galaxies GAL 26
Cannon, Annie Jump (US astronomer) TODAY 29
Canterbury (England) MOON 7
Canyons
  On Mars MARS 10-11, 20-21, 24; PROJ 6-7; RPS 22 (see also Kasei Vallis and Valles Marineris)
  On Titania UR 22
  On Venus VENUS 23
Cape Canaveral, Florida (NASA launch site) PILOT 12-13; RPS 7-11, 28

Cape Kennedy (*see* Cape Canaveral, Florida)
Capek, Karel (Czech author) SF 29 (*see also* R.U.R.)
Capricorn (constellation) GUIDE 19
Caracol (ancient observatory) ANCIENT 8
Carbon LIFE 4, 14; MARS 14, 25; STAR 14, 27
Carbon dioxide EARTH 14, 27; LIFE 16; MARS 12, 24; PILOT 20; VENUS 15, 17, 25, 27
Carme (moon of Jupiter) JUP 14-15, 29
Carter, Jimmy (US president) RPS 25
"Case of the UFOs, The" ("NOVA" episode) UFO 15
Cassini, Giovanni (Italian astronomer) SAT 12
Cassini Division SAT 12-13, 21, 27 (*see also* Saturn: Rings of)
Cassiopeia (constellation) GUIDE 8-9
Caster (star) STAR 13
Caucasus Mountains (USSR) TODAY 7
Centaurus (constellation) GAL 28-29; MYTH 17
Centaurus A (galaxy) GAL 22
Central Bureau for Astronomical Telegrams TODAY 19 (*see also* "Astrograms")
Ceres (Roman goddess of agriculture) ASTER 11; MYTH 11
Ceres (asteroid) ASTER 5, 6-7, 9, 11, 15, 16, 22, 28, 29; MYTH 11
 As a planet ASTER 6, 15 (*see also* Planetoids)
Cetus (constellation) GUIDE 14-15; MYTH 28
Chaco Canyon (New Mexico) ANCIENT 8
Challenger (space shuttle) GARB 12-13; PILOT 17, 26; RPS 15; SF 24-25; WSP 13, 15, 28, 29
 Explosion of PILOT 17, 26
Challenger Deep EARTH 10
Channels, river, on Mars MARS 7, 11 (*see also* Canyons: On Mars)
Charge cards SF 24, 28
Charon (mythological figure) MYTH 10-11; PLUTO 12-13
Charon (moon of Pluto) ASTER 14, 15; COL 17; MOON 13; MYTH 10-11; NEP 25; PLUTO 12-13, 14-15, 18-19, 23, 25, 26-27, 28-29
 As captured asteroid ASTER 14, 15 (*see also* Pluto-Charon system)
Cheesefoot Head, Hampshire (England) UFO 22
Cheyenne Mountain (US) GARB 11
Chicago, Illinois (US) EARTH 29
Chichén Itzá (Mexico) ANCIENT 8, 29
China ANCIENT 6-7, 17, 21; COMET 28; MYTH 4, 8, 19, 25; PILOT 5; QPB 12; RPS 4-5, 29; WSP 9, 29
China 1 (satellite) (*see* Tung-Fung-Hung)
Chiron (asteroid) ASTER 14, 28; SAT 23
 As escaped moon of Saturn ASTER 28; SAT 23
 As a planetesimal ASTER 28
Christy, James W. (US astronomer) PLUTO 13
Chromosphere SUN 6
Chumash Indians MYTH 29
Circinus (constellation) GUIDE 29; MYTH 21
Circle of Chiefs (American Indian constellation) MYTH 29
Cities in space EARTH 25 (*see also* Space colonies)
Civil War, US MYTH 15
Clark University (Worcester, Massachusetts) RPS 7
Clock, the (*see* Horologium)
Clocks, stellar QPB 14
Close Encounters of the First, Second, and Third Kinds UFO 28-29
Close Encounters of the Third Kind (movie) UFO 29
Clouds EARTH 14, 15, 18; GUIDE 21; JUP 12, 13, 25; NEP 4-5, 6-7, 14-15, 16-17, 18, 26; RPS 20-21; SAT 10-11; UFO 14; UR 18-19; VENUS 8-9, 10, 12-13, 15, 25, 28-29; WSP 8, 9
Clusters of galaxies (*see* Galactic clusters)
Coal Sack (dark nebula) GUIDE 28-29
Coelacanth (living fossil) DINOS 25
Colliding galaxies GAL 26-27; QPB 27
Collins, Michael (astronaut) MOON 16; PILOT 11 (*see also* Apollo missions: Apollo 11)
Colonies on other worlds (*see* Space colonies)
Colorado (US) DINOS 7; GARB 11
Colorado River RPS 15 (*see also* Grand Canyon)
Columbia (command module) MOON 16 (*see also* Apollo missions: Apollo 11)
Columbia (space shuttle) PILOT 13
Coma Berenices (constellation) GAL 16
Coma Berenices (galactic cluster) GAL 16
Comet Bennett COMET 9
Comet Halley (*see* Halley's Comet)
Comet Howard-Koomen COMET 22
Comet Ikeya-Seki COMET 4-5
Comet Kobayashi-Berger-Milon 1975h SOL 23
Comet Kohoutek COMET 17
"Comet pills" COMET 17
Comet Seki-Lines COMET 16-17
Comet West COMET 17; SOL 22
Comets ANCIENT 24; ASTER 16-17, 24; COMET 4-5, 7, 9-13, 16-23, 24-27, 28-29; DINOS 12-13, 14-15, 16, 18-19, 20, 22-23,

24-25, 26, 29; GUIDE 24; JUP 9; LIFE 9, 17; MERC 25; MYTH 14-15; NEP 20, 24-25; PILOT 25; PLUTO 15, 25, 27; PROJ 12, 16; SAT 19; SOL 22-23, 27; SUN 8; TODAY 19; UFO 14, 29; UR 5
  As omens of disaster ANCIENT 24; COMET 20-21, 28-29; DINOS 15; MYTH 14-15
  Orbits of ANCIENT 24; COMET 18, 28-29; MERC 25
Command modules WSP 6 (see also Apollo missions and Lunar module)
Communications satellites GARB 14, 29; RPS 5, 10-11, 14, 28-29; WSP 10-11, 29
Compass, magnetic ANCIENT 7; EARTH 16; PROJ 24; SUN 15; VENUS 20
Compasses, the (constellation) (see Circinus)
Computers SF 7, 9, 25, 26, 28
Confederate States of America (see Civil War, US)
Conrad, Charles (astronaut) MOON 17
Constantinople COMET 29
Constellation projector PROJ 22-23
Constellations ANCIENT 4, 6; GAL 28-29; MYTH 16-17, 19, 20-21, 22-23, 28-29; PROJ 16, 22
  Differing interpretations of MYTH 16 (Orion), 17 (Taurus), 29 (The Big Dipper)
  "Lost" MYTH 28-29
  "New" MYTH 20-21; PROJ 22
  (see also names of individual constellations and the book **The Space Spotter's Guide**)
Continents
  Of Earth EARTH 8; VENUS 22-23 (see also Africa, Antarctica, Asia, Australia, Europe, North America, and South America)
  Of Venus VENUS 21, 22-23 (see also Aphrodite Terra and Ishtar Terra)
Copernican system ANCIENT 22-23, 24, 27; UNIV 6-7
Copernicus, Nicolaus (Polish astronomer) ANCIENT 22-23, 24, 27; MERC 21; UNIV 6-7
Cores (of planets) (see subentries under Earth, Jupiter, Mercury, Moon, Saturn, Uranus, and Venus)
Corona MOON 11; SUN 6, 16, 17, 20, 22, 24-25, 28
Coronagraphs SUN 17, 22
Coronal holes SUN 6, 24-25
"Cosmic pizza" SOL 7 (see also Solar system)
"Cosmic Quest" (game) LIFE 16-17
Cosmic rays GARB 21; MOON 22, 25; WSP 21

Cosmonauts (see Astronauts and cosmonauts)
Cosmos 954 (satellite) GARB 16-17
Crab Nebula ANCIENT 21; QPB 12-13; UNIV 25
Crab pulsar QPB 12-13
Cradle Hill (England) UFO 13
Craters ANCIENT 27; ASTER 18; COMET 6-7, 24-25; DINOS 8-9, 14, 16-17, 18; EARTH 6-7; GARB 12-13, 26; GUIDE 21; JUP 16-17, 18-19, 21, 23; MARS 8-9, 10, 16-17, 20, 24; MERC 4-5, 10-11, 12-13, 15, 28-29; MOON 6, 7, 14, 19, 28-29; NEP 10-11; PROJ 8-9, 20; RPS 17, 18-19, 22, 25; SAT 20-21; TODAY 23; UR 21, 22-23, 25; VENUS 21 (see also Barringer Meteor Crater, Caloris basin, Langrenus, Nova Scotia crater, Stickney, and Wolf Creek Site)
Cronus (Greek god) SAT 5 (see also Saturn [Roman god of agriculture])
Crows LIFE 22
Crust, Earth's EARTH 7, 8, 10; GARB 14 (see also Plates)
Crux (constellation) (see Southern Cross)
Cuba RPS 13; WSP 29
Cygnus (constellation) GUIDE 12-13; QPB 19, 21
Cyrano de Bergerac (see Bergerac, Cyrano de)
Czechoslovakia WSP 29

D

Daguerrotype (early photograph) MOON 4
Dark Ages ANCIENT 20-21
Dark nebulas PROJ 20
D'Arrest, Heinrich L. (German astronomer) NEP 5
Darth Vader (fictional character) SF 26 (see also Star Wars [movie])
Davis, Jefferson (Confederate president) MYTH 15
Davis, Raymond (US scientist) SUN 7
de Bergerac, Cyrano (see Bergerac, Cyrano de)
Democritus (Greek philosopher) MYTH 16
"Demon Star" (see Algol)
Deimos (moon of Mars) ASTER 8-9; MARS 16, 25, 28-29; SOL 18
  As captured asteroid ASTER 8-9; MARS 25
Dendera (see Zodiac of Dendera)
Deneb (star) GUIDE 12-13; MYTH 25
  As a pole star MYTH 25
Denmark ANCIENT 29
Deserts, Martian MARS 6
Deuterium (see Hydrogen)

Devil, the VENUS 7
Devil's Punchbowl, the (England) UFO 22
Dinosaurs COMET 25; SF 14; UFO 24; VENUS 10 (*see also the book* **Did Comets Kill the Dinosaurs?**)
Dione (moon of Saturn) SAT 13, 16-17, 22, 29
Dirigibles (airships) COL 14-15; SF 24
Discovery (space shuttle) PILOT 12-13; RPS 10; WSP 29
Disintegrator rays SF 26
"Dog Days" MYTH 5 (*see also* Sirius)
Donati's Comet COMET 16-17
Doppler, Christian (Austrian physicist) UNIV 10
Doppler effect (*see* Red shift *and* Violet shift)
Dorado (constellation) GUIDE 29
Double planet(s)
    Earth-Moon system as EARTH 29; MOON 12-13, 18; PLUTO 12-13
    Pluto-Charon system as ASTER 15; MOON 13; PLUTO 12-13, 19, 28-29
Double pulsars QPB 16
Double-star system(s) ANCIENT 11; GAL 10, 25; PROJ 20; QPB 10-11; SOL 11
    The Sun as possible part of DINOS 20-21, 22; SOL 11 (*see also* Nemesis)
    (*see also* Binary stars *and* Multiple stars)
Dragons MYTH 4, 13
Dubhe (star) GUIDE 8 (*see also* Pointers)
Dust storms, Martian MARS 6 (*see also* Weather: On Mars)
Dutch (*see* Holland and the Dutch)
Dwarf galaxies GAL 12, 23 (*see also* Large Magellanic Cloud *and* Small Magellanic Cloud)

E

Eagle (lunar module) MOON 16; PILOT 11 (*see also* Apollo missions: Apollo 11)
Eagle Nebula SOL 5
"Ears," planet with JUP 12; SAT 4-5 (*see also* Saturn)
Earth ANCIENT 6, 13, 14-15, 16, 18, 22-23, 24; ASTER 4, 6-7, 18-19, 20, 21, 22, 24, 25, 28; COL 5, 8, 10-11, 13, 15, 18, 21, 22-23, 25, 26-27, 28-29; COMET 6-7, 9, 11, 12, 15, 20, 22, 24-25, 27, 28-29; DINOS 5, 6, 8-9, 10-11, 12-13, 14, 18-19, 20, 22-23, 24-25, 26-27, 28-29; GAL 5, 7, 8, 11, 20, 26, 28-29; GARB 4-5, 6, 7, 8, 11, 12, 14-15, 16-17, 19, 21, 24-25, 26-27, 28-29; GUIDE 5, 6, 16, 18-19, 23, 25, 26; JUP 4, 9, 10-11, 12-13, 16-17, 24-25, 27, 28-29; LIFE 4-5, 6-7, 8, 9, 10, 12-13, 16-17, 18, 19, 21, 22, 24-25, 26-27; MARS 4, 9, 10, 12, 13, 14, 16, 18, 20, 22, 24, 25, 27, 28-29; MERC 5, 7, 8, 9, 10, 14-15, 18-19, 20-21, 23, 28-29; MOON 4, 7, 8-9, 10-11, 12-13, 14-15, 16, 18, 20-21, 22, 23, 24-25, 26-27, 28-29; MYTH 5, 12-13, 14, 16, 17, 18, 21, 23, 25, 28; NEP 4-5, 6-7, 9, 12-13, 16-17, 19, 20, 24-25, 26-27, 28-29; PILOT 5, 7, 15, 17, 23, 24-25, 26, 27, 28; PLUTO 9, 12-13, 14-15, 16-17, 18, 19, 21, 23, 27, 28-29; PROJ 4, 6, 8, 10, 11, 12, 13, 16, 20, 22, 28; QPB 6-7, 10, 14, 24; RPS 4, 6, 7, 8-9, 10-11, 12-13, 14-15, 16, 18-19, 20, 23, 25, 26-27, 28-29; SAT 6-7, 8-9, 14, 16, 25, 28-29; SF 7, 12-13, 14, 16, 19, 20, 22, 24-25; SOL 8, 10, 12-13, 18, 19, 20, 21, 22, 24-25; STAR 10, 13, 20, 22-23, 24, 27, 29; SUN 8, 9, 10-11, 12, 14-15, 17, 18, 19, 20-21, 24, 26, 28-29; TODAY 7, 8-9, 13, 15, 17, 25, 27; UFO 17, 19, 20, 27, 29; UNIV 5, 6-7, 14-15, 22-23, 24, 26, 28-29; UR 6, 8, 10, 11, 17, 18, 20, 22, 28-29; VENUS 7, 9, 10-11, 12-13, 15, 16-17, 19, 20, 25, 27, 28-29; WSP 6, 7, 8, 9, 10-11, 12, 13, 15, 17, 19, 21, 22, 23, 25, 26-27
    Atmosphere of ASTER 18-19; COL 5, 28; COMET 7, 9, 12, 24-25; EARTH 6-7, 14-15, 16-17, 20, 22-23; GARB 8, 12, 16-17, 19, 28, 29; LIFE 6, 16-17; MARS 9; PILOT 5, 15, 28; PROJ 16, 18; RPS 7, 12-13; SAT 25; SOL 20; SUN 18, 24; TODAY 7, 9, 17; WSP 6, 9, 12, 15, 19, 21, 22-23, 26-27
    Axial tilt of GAL 28; MARS 4; NEP 19; SAT 28; SUN 9; UR 8
    Continents of EARTH 8; VENUS 22-23 (*see also* Africa, Antarctica, Asia, Australasia, Australia, Europe, North America, *and* South America)
    Core of EARTH 7, 16, 23; MERC 8; MOON 23; VENUS 12-13, 20
    Equator of GAL 12-13; RPS 28; SAT 7
    Eventual destruction of SOL 24-25; STAR 22-23, 29
    Magnetic field of (magnetosphere) EARTH 16-17; JUP 25; NEP 19; SUN 19; VENUS 20
    Mountains of VENUS 22-23 (*see also* names of *individual mountains and mountain ranges*)
    Oceans of DINOS 16-17; EARTH 6, 8-9, 10, 15, 20-21, 22; VENUS 17, 25 (*see also* Atlantic Ocean *and* Pacific Ocean)
    Orbit of JUP 29; MARS 29; MERC 29; NEP 29; PLUTO 29; PROJ 13; SAT 29; SUN 9; UR 29; VENUS 11, 29
    Poles of

(see North Pole: Of Earth and South Pole)
Rotation of EARTH 16; JUP 9, 28; MARS 29; MERC 28; MOON 23, 25; NEP 28; PLUTO 28; RPS 20; SAT 28; UR 29; VENUS 28
(see also the book **Earth: Our Home Base**)
"Earth-grazers" (asteroids) ASTER 18-19
Earth-Moon system MOON 12-13, 18; PLUTO 12-13; PROJ 10-11; SUN 9 (see also Double planet[s]: Earth-Moon system as)
"Earthlings" MOON 14, 16; UFO 4-5, 9
Earthquakes EARTH 10-11, 12-13; GARB 14; VENUS 22
East Germany WSP 29 (see also Germany and the Germans)
Easter MOON 20
Eclipses ANCIENT 6-7, 9, 15, 18; GUIDE 6, 24; MERC 25; MOON 10-11; MYTH 12-13; PLUTO 18-19, 26-27; PROJ 10, 20; SUN 20-21, 22; TODAY 21
   Artificial SUN 22 (see also Coronagraphs)
   Lunar ANCIENT 6, 18; GUIDE 6; MOON 10-11; MYTH 12, 13; PROJ 10, 20
   Of Pluto-Charon PLUTO 18-19, 26-27
   Solar ANCIENT 6, 15; MERC 25; MOON 10-11; MYTH 12; PROJ 10; SUN 20-21, 22; TODAY 21
   Of a star NEP 13; PLUTO 14; UR 12-13
Ecosphere SOL 10
Egyptians, ancient ANCIENT 10-11, 16, 18, 28; GUIDE 11, 16; MYTH 4-5, 6, 19; SUN 10-11; UFO 8; UNIV 4
Einstein, Albert (German physicist) MERC 24-25; TODAY 24-25
Einstein ring TODAY 25
Elara (moon of Jupiter) JUP 14-15, 29
Electron (atomic particle) QPB 16
Elijah (Biblical prophet) UFO 6 (see also Bible)
Ellipses PLUTO 11 (see also Orbits)
Elliptical galaxies GAL 18, 22-23
Enceladus (moon of Saturn) SAT 16-17, 18, 28-29
England and the English (see Britain and the British)
English (language) MARS 20
English literature UR 10
Epimetheus (moon of Saturn) SAT 16-17, 28-29
Equator (see subentries under Earth, Pluto, and Saturn)
Eratosthenes (Greek astronomer) ANCIENT 18-19
Eros (Greek god) MYTH 9, 11
Eros (asteroid) ASTER 28, 29; MYTH 11

Erosion PROJ 6-7, 8
Eruptive prominences (see Prominences)
ESA (see European Space Agency)
Eskimo Nebula STAR 25
ESP (extra-sensory perception) (see Psychokinesis)
Ethane (gas) LIFE 4
Europa (Galilean moon of Jupiter) JUP 14-15, 20-21, 28-29; LIFE 16, 18-19; PILOT 11; RPS 24-25; SOL 13, 19
Europe and Europeans ANCIENT 7, 9, 20-21, 23, 24; COMET 28; EARTH 8; GAL 12; GARB 14; MYTH 4; RPS 4, 15; UFO 8-9; UR 11
European Space Agency (ESA) COMET 27; PILOT 13, 27; RPS 4, 16, 28-29; WSP 9, 11, 13, 28, 29
Evening Star (see also Morning Star and Venus) ANCIENT 15; GUIDE 21; VENUS 5, 28
Everest, Mt. EARTH 10; GAL 29; MARS 13
Everglades (Florida) RPS 13
Expanding Universe, concept of QPB 24; UNIV 16-21, 26-27 (see also Big Bang)
Exploding galaxies GAL 24-25
Exploding stars DINOS 29 (see also Novas and Supernovas)
Explorer (research satellites)
   Explorer 1 (first US satellite) RPS 8; WSP 28
   Explorer 42 (black hole detector) TODAY 29
Extraterrestrials (see Alien life)
Ezekiel (Biblical prophet) UFO 6-7 (see also Bible)

F

Faris, Mohammed (cosmonaut) WSP 29
Farkas, Bertalan (cosmonaut) WSP 29
Father Sky (American Indian god) MYTH 23
Faults EARTH 10-11, 13
"Federation, Galactic" (see "Galactic Empire")
Felis (former constellation) MYTH 29
Films, science fiction SF 4-5, 12, 19, 25, 26 (see also titles of individual films)
Fire Star (ancient Chinese name for Mars) MYTH 8 (see also Mars [planet])
Fireballs COMET 6-7 (see also Meteors)
Fireworks WSP 5
First Men in the Moon (movie) SF 12
First Men in the Moon, The (novel) SF 4, 12, 29
Fish DINOS 25; EARTH 21; LIFE 22 (see also Coelacanth and Lemon Butterfly fish)
Fisher, William F. (astronaut) RPS 10
Five Weeks in a Balloon (novel) SF 5

Fizeau, Armand H. L. (French scientist) TODAY 29
Flamsteed, John (British astronomer) UR 13
Flat Earth theory ANCIENT 15; UNIV 4-5
Flora (asteroid) ASTER 29
Flores (asteroid family) ASTER 29
Florida (US) RPS 13, 15, 28; SF 4; WSP 5 (*see also* Cape Canaveral, Florida)
Flying saucers SF 6-7; UFO 4-5, 10, 16-17, 18-19, 22-23, 28
Focal length PROJ 18-19
Fog on Mars MARS 21 (*see also* Weather: On Mars)
Follini, Stefania (Italian researcher) COL 11
Foo-fighters UFO 8-9
Fossils COL 5; DINOS 6-7
France and the French ASTER 15; COMET 16-17; NEP 5; RPS 28-29; SF 4; TODAY 29; WSP 28
Frankenstein (novel) SF 5
Fraunhofer, Joseph von (German optician) TODAY 29
Freedom (space station) SF 16
French Guiana RPS 28
From the Earth to the Moon (novel) SF 4, 29 (*see also* Verne, Jules)
Fuji, Mt. MARS 13
Fusion (*see* Nuclear fusion)

G

G-force PILOT 18; WSP 15
Gagarin, Yuri (cosmonaut) PILOT 4-5, 13; WSP 5, 28
Galactic clusters GAL 14, 16-17, 22, 26; UNIV 14-15, 16-17 (*see also* Coma Berenices [galactic cluster], Local Group, *and* Virgo cluster)
Galactic cores GAL 7, 9, 18-19, 20, 22, 24-25; QPB 20-21, 22-23, 26, 28-29
"Galactic Empire" (fictional future government) SF 19, 20, 22 (*see also Star Wars* [movie])
Galactic midline DINOS 22-23, 24
Galaxies COL 18-19; DINOS 22-23; GAL 9, 12-13, 14-15, 16-17, 18, 20-21, 22-23, 24-25, 26-27; GUIDE 14-15, 26-27, 28-29; LIFE 20-21, 26; PROJ 20; QPB 8, 20-21, 22-23, 24-25, 26-27; SF 28; SOL 23, 26-27; STAR 5, 6, 17, 18-19; TODAY 10-11, 15, 25, 28, 29; UFO 27; UNIV 8-9, 10-11, 12-13, 14-15, 16-17, 18-19, 20-21, 22, 24-25, 26-27, 29 (*see also* Andromeda Galaxy, Milky Way Galaxy, *other specific galaxies and types of galaxies, and the book* **Our Milky Way and Other Galaxies**)
Galaxies, primal QPB 25, 29
Galilean moons of Jupiter ANCIENT 27; GUIDE 23; JUP 4-5, 6-7, 14-15, 16-23; MOON 12; NEP 20-21; RPS 25; SOL 18-19; UR 22 (*see also* Callisto, Europa, Ganymede, *and* Io)
Galileo Galilei (Italian astronomer) ANCIENT 26-27; JUP 4-5, 7, 12, 15; MOON 6-7, 17; PROJ 18, 26; SAT 4-5; SUN 27; VENUS 7, 9
Galileo (probe) JUP 26-27; RPS 16, 26
Galle, Johann Gottfried (German astronomer) NEP 5, 24-25
Game (*see* "Cosmic Quest")
Gamma rays TODAY 28
Ganymede (Galilean moon of Jupiter) JUP 14-15, 18-19, 28-29; MERC 26; PLUTO 12; RPS 25, 27; SAT 25; SOL 13, 19
Gardner, Dale (astronaut) RPS 10
Garn, Jake (politician and astronaut) PILOT 14-15
Garneau, Marc (astronaut) WSP 28
Gas clouds, stellar STAR 5-11, 16-17, 27 (*see also* Protostars)
Gas giants (planets) COL 17; JUP 27; NEP 4-5, 7, 12, 19; PILOT 27; PROJ 12; SAT 6-7, 10; SOL 8, 14-15, 16, 18; UR 6, 17; WSP 21 (*see also planet entries* Jupiter, Neptune, Saturn, *and* Uranus)
Gas jets GAL 25; QPB 20-21, 28-29
Gasoline RPS 6
Gauss, Karl (German mathematician) ASTER 9
Gemini (constellation) GUIDE 19; MYTH 18; PROJ 16; STAR 25
Gemini missions PILOT 6-7, 9
Geminids (meteor shower) PROJ 16
George III (king of Britain) UR 11
"George's Star" (*see* Georgium Sidus)
Georgium Sidus ("George's Star") UR 11 (*see also* Uranus [planet])
Geostationary satellites RPS 14-15
Germany and the Germans COMET 28; RPS 8; TODAY 29; UFO 6; WSP 5, 7, 15, 28, 29
Gernsback, Hugo (US author and editor) SF 6-7, 28-29 (*see also Amazing Stories* and *Ralph 124C 41+*)
Geysers ASTER 24; NEP 22-23 (*see also* "Ice volcanoes")
"Ghoul, the" (*see* Algol)
Giant planets (*see* Gas giants)
Giotto (Italian painter) COMET 29
Giotto (probe) COMET 19, 26-27
Glaciers EARTH 21; JUP 21; MARS 24; RPS 24-25

Glauke (asteroid) *ASTER* 28
Glenn, John (astronaut) *PILOT* 5
"Global winter" *DINOS* 10-11, 29
Globular clusters *GAL* 10-11
God *VENUS* 7
Goddard, Robert H. (US rocket designer) *RPS* 6-7; *WSP* 5, 7
Goddess of liberty (*see* "Mercury" dime)
Goddesses of love (*see* Aphrodite, Ishtar, *and* Venus [planet])
Godfrey, Alan (alleged UFO abductee) *UFO* 20
Godwin, Francis (British author) *SF* 8 (*see also* "Man in the Moon")
Gold Star (ancient Chinese name for Venus) *MYTH* 8 (*see also* Venus [planet])
Golden Gate Bridge (US) *DINOS* 27
Golden Peg (ancient Mongolian name for the North Star) *MYTH* 25
Golf balls *GARB* 5
Gosses Bluff (Australia) *DINOS* 9
Grand Canyon (Arizona) *MARS* 10; *RPS* 15; *UR* 24
Granules *SUN* 6, 12-13
Gravitational lens *TODAY* 25
Gravity and gravitational pull *ASTER* 8; *COL* 7, 13, 15, 21, 28-29; *COMET* 18, 29; *DINOS* 12-13, 18-19, 20, 23; *EARTH* 4; *GAL* 9, 14, 17, 24-25; *JUP* 12, 23, 29; *MARS* 16, 22, 29; *MERC* 5, 8, 25, 29; *MOON* 4, 8, 17, 18, 21, 22, 25, 29; *NEP* 5, 6, 9, 11, 20; *PILOT* 14-15, 18-19, 22, 23, 24-25; *PLUTO* 4-5, 9, 19, 23, 25, 29; *QPB* 4, 6, 9, 10-11, 14-15, 16, 21, 27; *SAT* 7, 15, 21, 22-23, 29; *SF* 4, 10-11, 16; *SOL* 4, 8, 12, 14, 17; *STAR* 6-7, 10, 17, 20, 21; *SUN* 4, 6, 8, 11, 29; *TODAY* 24-25; *UNIV* 16, 26; *UR* 26, 29; *VENUS* 17; *WSP* 15, 17
Gravity, artificial *COL* 7, 28; *PILOT* 14, 24-25; *SF* 10-11, 16; *WSP* 17, 18
Gravity, zero (*see* Zero gravity)
Great Bear (*see* Ursa Major)
Great Britain (*see* Britain and the British)
Great Dark Spot *NEP* 4-5, 6, 14-15, 16-17, 18
Great Dog (*see* Canis Major)
Great Galaxy (*see* Andromeda Galaxy)
Great Moon Hoax *LIFE* 10-11, 16; *UFO* 24 (*see also* Locke, Richard B. *and* New York *Sun*)
Great Nebula (*see* Orion Nebula)
Great Pyramid (*see* Pyramid, Great)
Great Red Spot *EARTH* 18-19; *JUP* 10-11, 13, 25; *NEP* 16-17
Great War Comet of 1861 *MYTH* 15
Greeks, ancient *ANCIENT* 14-19, 20-21, 23, 24, 27, *ASTER* 12; *COMET* 28; *GUIDE* 28; *MYTH* 5, 6-7, 8, 10, 16-17, 19, 25, 28; *NEP* 9; *UNIV* 4, 6; *VENUS* 7
"Greenhouse effect" *VENUS* 26-27
Ground stations *WSP* 10-11
"Guest Cosmonaut" program (*see* Intercosmos)
Gulf of Mexico *RPS* 13
Gulliver's Travels (novel) *SF* 14 (*see also* Swift, Jonathan)
Gum Nebula *ANCIENT* 23
Gunpowder *RPS* 4, 6
Gurragcha, Jugderdemidyin (cosmonaut) *WSP* 29

H

"Hairy stars" *COMET* 5, 20, 28 (*see also* Comets)
Hale, Edward Everett (US author) *SF* 28
Hall, Asaph (US astronomer) *MARS* 16; *TODAY* 22-23
Halley, Edmund (British astronomer) *ANCIENT* 24; *COMET* 18-10, 28-29; *UNIV* 6
Halley's Comet *COMET* 17, 18-21, 26-27, 28-29; *DINOS* 19; *MYTH* 14; *PILOT* 27, 28
As an omen of disaster *COMET* 20-21, 28-29; *MYTH* 14
Ham (chimp in space) *PILOT* 5
Harold (king of England) *COMET* 28
Harrington, Bob (US astronomer) *PLUTO* 25
Harvard Observatory *TODAY* 23
Hathor (asteroid) *ASTER* 28
Hawaiian Islands (US) *ANCIENT* 29; *EARTH* 8-9, 12-13, 21; *SOL* 26; *TODAY* 6-7; *UFO* 10
Hawking, Stephen (British physicist) *QPB* 19
Hazard, Cyril (Australian astronomer) *TODAY* 29
Heat (*see* Infrared radiation)
Hecates Tholus (inactive Martian volcano) *MARS* 25
Hector (asteroid) *ASTER* 10, 11, 29
Heinlein, Robert (US author) *SF* 7
Helin, Eleanor (US astronomer) *NEP* 24-25
Helios (Greek god of the Sun) *MYTH* 5; *UNIV* 4
Helium *JUP* 9, 27, 28; *LIFE* 14, 17; *QPB* 6-7; *SAT* 8-9, 10; *SOL* 6, 12, 14, 29; *STAR* 5, 8, 10, 14, 22, 27, 28-29; *SUN* 6-7; *UR* 17
Helix Nebula *GUIDE* 27
Hendry, Allan (UFO investigator) *UFO* 15
Hercules (constellation) *GUIDE* 26
Herculina (asteroid) *ASTER* 28, 29
Hermaszewski, Miroslaw (cosmonaut) *WSP* 29
Hermes (mini-shuttle) *PILOT* 13
Herschel, Caroline (German-born British astronomer) *NEP* 24-25

Herschel, William (German-born British astronomer) MYTH 29; NEP 24-25; TODAY 29; UNIV 8; UR 5, 10-11, 16
Herschel (proposed name for the planet Uranus) UR 11 (*see also* Uranus [planet])
Hertz, Heinrich Rudolph (German scientist) TODAY 29
Hidalgo (asteroid) ASTER 14, 28
Hill, John (British eccentric) MYTH 28
Himalayas (mountains) EARTH 10; VENUS 22
Himalia (moon of Jupiter) JUP 14-15, 28-29
Hindus, ancient MYTH 13 (*see also* India, ancient)
Hipparchus (Greek astronomer) ANCIENT 17
Hiroshima (Japan) QPB 11
HL-10 Lifting Body PILOT 12-13
Hoaxes (*see* Great Moon Hoax *and* UFOs)
Holland and the Dutch ANCIENT 27; PROJ 18; WSP 29
Homer (Greek author) SF 5 (*see also* Odyssey, The)
Horologium (constellation) GUIDE 29; MYTH 20-21
Horoscope MYTH 26-27 (*see also* Astrology)
HOTOL (**Ho**rizontal **T**ake-off and **L**anding Craft) WSP 14
Houston, Texas (US) MOON 7; SF 13
Hubble Space Telescope ANCIENT 27, 29; GARB 28-29; PLUTO 23; RPS 16, 27; TODAY 8-9; WSP 23
Hungary WSP 29
Hurricanes EARTH 15; GARB 15; RPS 13; WSP 8
    Hurricane Gladys RPS 13
    Hurricane Juan GARB 15
    (*see also* Great Dark Spot *and* Great Red Spot)
Huxley, Aldous (British author) SF 29 (*see also* Brave New World)
Huygens, Christian (Dutch astronomer) SAT 5
Hydrogen COL 20; JUP 8-9, 27, 28; LIFE 4, 14, 17; QPB 6-7, 8, 12, 24; RPS 5, 6; SAT 8-9, 10; SOL 6, 12, 14, 24, 29; STAR 5, 8, 10, 12, 14, 27, 28-29; SUN 6-7, 11; UR 17
Hydrogen bomb SUN 6-7 (*see also* Atomic bomb and atomic energy *and* Nuclear fusion)
Hynek, J. Allen (UFO researcher) UFO 22, 27, 28-29
Hyperdrive LIFE 17
Hyperion (moon of Saturn) SAT 16-17, 29
Hyperspace SF 20

I

Iapetus (moon of Saturn) SAT 16-17, 18-19, 29
IAU (*see* International Astronomical Union)
Icarus (asteroid) ASTER 28; MERC 25
Ice ages EARTH 21
Ice caps of Mars (*see* Polar ice caps: Of Mars)
"Ice volcanoes" JUP 21; NEP 22-23
Impact basins (*see* Caloris basin *and* Craters)
In the Twenty-Ninth Century — The Day of an American Journalist (novel) SF 28-29 (*see also* Verne, Jules)
Incas MYTH 22-23
India EARTH 10; RPS 29; SUN 24; WSP 9, 29
India, ancient MYTH 13, 25 (*see also* Hindus, ancient)
Indians, American ANCIENT 8-9, 21, 28-29; MYTH 6, 16, 19, 22-23, 28-29; QPB 12; SUN 22; UFO 9 (*see also* names of individual tribes and groups)
Indonesia EARTH 13; RPS 11; WSP 29
Industrial Revolution SF 4
Infrared Astronomical Satellite (*see* IRAS)
Infrared radiation TODAY 4-5, 10-11, 28-29; VENUS 13, 26-27
Infrared Space Observatory (space telescope) GARB 14
Inner planets (*see* Earth, Mars, Mercury, *and* Venus)
Insects COL 9, 29; LIFE 22
Intasat (satellite) WSP 29
Intercosmos ("Guest Cosmonaut" program) WSP 28-29
International Astronomical Union (IAU) MOON 14; VENUS 23
International Astronomical Union Crater No. 221 (lunar crater) MOON 14
Io (mythological figure) MYTH 11
Io (Galilean moon of Jupiter) EARTH 18-19; JUP 14-15, 22-23, 24-25, 28-29; MYTH 11; RPS 24; SOL 13, 19
    Orbit of JUP 14-15, 23, 24-25, 28
    Volcanoes on EARTH 3, 18-19; JUP 22-23, 24-25; RPS 24
Ion drives COL 20, 23
IRAS (satellite) PLUTO 24-25; TODAY 11
Iridium COMET 25; DINOS 8
Iron STAR 14, 17, 27; VENUS 20
Iron, meteoric (*see* Meteoric iron)
Iron, molten VENUS 20
Irwin, Jim (astronaut) MOON 17
Isaiah (Biblical prophet) VENUS 7 (*see also* Bible)
Ishtar (Babylonian goddess of love) VENUS 23
Ishtar Terra ("continent" on Venus) VENUS 22-23

Isis (ancient Egyptian goddess) MYTH 6
Islam (*see* Arabs, ancient, Black Stone of the *Kaaba*, *and* Muslims)
"Island universes" (galaxies) GAL 15
Islands as mountains EARTH 8-9
ISO (*see* Infrared Space Observatory)
Italy and the Italians PROJ 18, 26; RPS 15, 29; VENUS 7; WSP 19, 28
Ivanov, Georgi Ivan (cosmonaut) WSP 29

J

Jackson Lake (Wyoming) ASTER 19
Jahn, Sigmund (cosmonaut) WSP 29
Jansky, Karl Guthe (US engineer) TODAY 29
Janus (moon of Saturn) SAT 16-17, 28-29
Japan COMET 27; RPS 4, 11, 28-29; SUN 24; WSP 15, 29
Japanese, ancient ANCIENT 21; MYTH 16, 19
Jerusalem COMET 28
Jews, ancient COMET 21, 28
Juno (Roman goddess) ASTER 11; MYTH 11
Juno (asteroid) ASTER 11, 29; MYTH 11
Jupiter (Roman god of justice) MYTH 8; UR 11
Jupiter (planet) ANCIENT 13, 16, 27; ASTER 4, 5, 6-7, 12-13, 14, 16, 25, 26-27; COL 11, 17; COMET 29; EARTH 18-19, 28-29; GARB 26; GUIDE 18, 22-23; LIFE 14, 16-17, 18; MARS 16, 28-29; MERC 26, 29; MOON 12; MYTH 8, 11; NEP 7, 12, 15, 16-17, 19, 28-29; PILOT 11, 25; PLUTO 12, 23, 28-29; PROJ 12, 13, 18, 20; RPS 24-25, 26-27; SAT 6-7, 10, 13, 28-29; SOL 8-9, 10, 13, 14-15, 18-19, 20-21; STAR 27; SUN 29; TODAY 23; UNIV 6, 29; UR 6, 11, 14, 16-17, 18, 22, 28-29; VENUS 28-29; WSP 21
    Atmosphere of JUP 8-9, 10, 24-25, 26-27; LIFE 14, 16; RPS 26; UR 18
    Belts of JUP 10; UR 18
    Clouds of JUP 12, 13, 25; UR 18
    Colors of JUP 25
    Core of JUP 8
    Magnetic field of JUP 24-25, 27; MERC 9; NEP 19
    Moons of (*see* Moons: Of Jupiter)
    Orbit of JUP 6, 29; PROJ 13
    Ring of JUP 12-13; NEP 12; RPS 25; SAT 13; SOL 21
    Rotation of JUP 9, 28
    Storms on EARTH 18-19; JUP 10-11, 13; NEP 17 (*see also* Great Red Spot)
    Wind on NEP 15
    Zones of JUP 10
    (*see also* the book **Jupiter: The Spotted Giant**)
Jupiter system, the JUP 14-15 (*see also* Moons: Of Jupiter)

K

Kaaba (sacred shrine of Islam) (*see* Black Stone of the Kaaba)
Kangaroo COL 26
Kant, Immanuel (German philosopher) GAL 15
Kasei Vallis (Martian canyon) MARS 21
Keck Telescope (Hawaii) TODAY 6-7
Kennedy, John F. (US president) PILOT 9
Kennedy Space Center (Florida) RPS 28; WSP 5 (*see also* Cape Canaveral, Florida)
Kenya RPS 29; TODAY 29
Kepler, Johannes (German astronomer) SF 14
Key, Francis Scott (US composer) RPS 5
Kilauea Crater, Hawaii (volcano) EARTH 12-13
Kirchoff, Gustav Robert (German scientist) TODAY 29
Kitt Peak Observatory (Arizona) MYTH 27; SOL 20; SUN 18, 22
Koala COL 26
Kowal, Charles (US astronomer) SAT 23
Krakatau (Krakatoa), Indonesia (volcano) EARTH 13
Kriege, David (amateur astronomer) TODAY 20-21
Kuiper Airborne Observatory PLUTO 14
Kulik, Leonid (Soviet scientist) DINOS 15

L

Lacaille, Nicholas Louis (French astronomer) GUIDE 29
LAGEOS (satellite) GARB 14, 27
Laika (dog in space) RPS 9
Lake Hoare (Antarctica) LIFE 22
Lake Okeechobee (Florida) RPS 13
Lakshmi Planum (plain on Venus) VENUS 22
Lalande, J. J. L. de (French astronomer) MYTH 29
Lambda Cephei (star) STAR 12
Landsat (satellite) RPS 14-15
Langrenus (lunar crater) MOON 7
Large Magellanic Cloud (galaxy) GAL 9, 12-13; GUIDE 28-29; STAR 18; TODAY 15
Lasers COL 20; GARB 6, 14; SF 24, 26, 28
Lava (*see* Molten rock)
Law of action and reaction RPS 4; SF 8
Law of gravity COMET 18, 29

Lead PILOT 27; VENUS 15
Lebanon RPS 15
Leda (mythological figure) MYTH 11
Leda (moon of Jupiter) JUP 14-15, 28; MYTH 11
Lemon Butterfly fish EARTH 21
Lenticular clouds UFO 14
Leo (constellation) GUIDE 10-11, 19; PROJ 16
Leonids (meteor shower) ASTER 19; PROJ 16; SOL 20-21
Leonov, Alexei (cosmonaut) PILOT 7
Leverrier, Urbain Jean Joseph (French astronomer) NEP 5, 25
LGM (see "Little Green Men")
Libra (constellation) GUIDE 19; MYTH 9, 28
Lichen LIFE 22-23
Life, alien or extraterrestrial (see Alien life)
Life, origins of EARTH 14, 20; LIFE 4, 6-7, 9; MARS 25 (see also Building blocks of life)
Lightning EARTH 15; JUP 24-26; VENUS 14-15, 17, 25
Light-years COL 18-19; GAL 7, 11, 12, 14, 16, 25, 26, 28; GARB 25; LIFE 8, 24, 26; NEP 27; QPB 12, 22, 25, 26; SOL 28; STAR 19; TODAY 15, 29; UFO 18; UNIV 8-9, 12, 14-15, 22-23, 25
Limax (former constellation) MYTH 28
Little Dipper (constellation) MYTH 23 (see also North Star)
"Little Green Men" (LGM) QPB 13; TODAY 11 (see also Alien life, Pulsars, and Radio signals or messages)
Lizards DINOS 5
Local Group (the Milky Way's galactic cluster) GAL 14, 16; UNIV 14-15
Locke, Richard B. (US journalist) UFO 24 (see also Great Moon Hoax and New York Sun)
London (England) RPS 8
Looking Backward, 2000-1887 (novel) SF 28 (see also Bellamy, Edward)
Loop prominences (see Prominences)
Los Angeles (California) SF 8
Lowell, Percival (US astronomer) LIFE 10; MARS 6-7, 11; MYTH 10-11; PLUTO 6-7
Lowell Observatory MYTH 11; PLUTO 7; TODAY 23
Lucifer (Biblical figure) VENUS 7
Luna (Soviet probes) MOON 15; PILOT 8-9
Lunar colonies (see Space colonies)
Lunar eclipses (see Eclipses: Lunar)
Lunar missions (see Apollo missions)
Lunar module WSP 6 (see also Apollo missions and Eagle)
Lunar orbiters MOON 14; PILOT 8-9

Lunar Rover COL 15 (see also Apollo missions)
Lunokhod (Soviet lunar rover) MARS 18; PILOT 27
Lyra (constellation) GUIDE 12-13; PROJ 16; STAR 24
Lyrids (meteor shower) PROJ 16
Lysithea (moon of Jupiter) JUP 14-15, 29

M

M31 (see Andromeda Galaxy)
M51 (see Whirlpool Galaxy)
M81 (galaxy) GAL 21
M82 (galaxy) GAL 24 (see also Exploding galaxies)
Magellan, Ferdinand (Spanish explorer) GAL 12
Magellan (probe) VENUS 23
Magellanic Clouds (galaxies) COL 19; GAL 12-13; LIFE 26; TODAY 15 (see also Large Magellanic Cloud *and* Small Magellanic Cloud)
Magnetic fields EARTH 16-17; JUP 24-25; MERC 9; MOON 22, 23; NEP 19; SUN 14, 16, 19; VENUS 20 (see also subentries for *individual planets*)
Magnetosphere (see Earth: Magnetic field of)
Magnifying glass PROJ 18, 26
Main sequence stars GAL 8; SOL 29; STAR 14
Mammals DINOS 5, 25
"Man in the Moon" COMET 20; MOON 4-5; MYTH 6-7; SF 5
Man in the Moon, The (novel) SF 8 (see also Godwin, Francis)
Manhattan COMET 27
Manhattan Project SF 7 (see also Atomic bomb and atomic energy)
Manned Maneuvering Unit (see MMU)
Mantle, Earth's EARTH 7
Marcellinus, Ammianus (pen name of Aaron Nadel) (see Nadel, Aaron)
Mare Moscoviense (lunar feature) MOON 14
Mare Orientale (lunar feature) MERC 10
Mariner probes GARB 28-29; LIFE 16; MARS 9, 10, 20; MERC 4-5, 14-15, 28-29; RPS 16, 18-19; VENUS 15, 18-19; WSP 21
  Mariner 2 VENUS 15, 18-19
  Mariner 4 MARS 9
  Mariner 9 MARS 10
  Mariner 10 LIFE 16; MERC 4-5, 14-15, 28-29; RPS 18-19
Marius, Simon (German astronomer) JUP 15
Mars (Roman god of war) MARS 3; MYTH 8; UR 11

Mars (planet) ANCIENT 13, 16; ASTER 4, 5, 6-7, 8-9, 11, 16, 22, 25, 29; COL 6, 8-9, 11, 14-15, 28-29; COMET 6-7; DINOS 26; EARTH 18-19, 21, 25, 27, 28-29; GARB 6, 20-21, 26; GUIDE 18, 22-23; JUP 29; LIFE 10, 12-13, 14, 15, 16-17, 22, 24; MERC 9, 16, 29; MYTH 8, 11; NEP 7, 28-29; PILOT 11, 16, 20-21, 22-23, 25, 26, 27; PLUTO 28-29; PROJ 4, 6-7, 8, 12; RPS 22-23, 24; SAT 28, 29; SF 7, 8, 14, 17; SOL 8, 10, 12-13, 18, 20-21; STAR 22, 29; SUN 21, 29; TODAY 22-23; UFO 17; UNIV 6; UR 11, 28-29; VENUS 10, 29; WSP 5, 19, 20-21
   Atmosphere of LIFE 17; MARS 6, 9, 12, 22-23, 26-27; PROJ 6; RPS 22; SF 14
   Axial tilt of MARS 4, 28
   Canyons of MARS 10-11, 20-21, 24; PROJ 6-7; RPS 22 (*see also* Kasei Vallis *and* Valles Marineris)
   Clouds of MARS 7
   Core of SOL 12
   Magnetic field, lack of MERC 9
   Moons of (*see* Deimos, Moons: Of Mars, *and* Phobos)
   Orbit of MARS 28-29
   Poles of MARS 6, 7, 24 (*see also* North Pole: Of Mars)
   Rotation of MARS 4, 18, 28-29
   Seasons on MARS 4, 7, 28
   Weather on MARS 10, 21
   (*see also the book* **Mars: Our Mysterious Neighbor**)
Mars mission (proposed US-USSR joint project) MARS 18-20
Mars Observer (proposed probe) WSP 21
Mars robot probe (automated car) GARB 20; MARS 18
"Marsquakes" PROJ 6
"Martians" (*see* Alien life)
Massachusetts (US) RPS 7; TODAY 19
Masum, M. D-G. (cosmonaut) WSP 29
Maunder, E. W. (British astronomer) SUN 27
Maunder minimum SUN 27 (*see also* Sunspot cycle)
Maxwell Montes (mountains on Venus) VENUS 22
Mayas ANCIENT 8-9
McAuliffe, Christa (schoolteacher and astronaut) PILOT 17
McDivitt, James A. (astronaut) MOON 16
McDonald Observatory (Texas) TODAY 26-27
McMath Solar Telescope (Kitt Peak) SUN 22
Medicine Wheel TODAY 5
Medusa (mythological figure) GAL 28-29; MYTH 25 (*see also* Algol)
Mendez, Arnaldo Tamayo (cosmonaut) WSP 29
Menger, Howard (alleged UFO abductee) UFO 20
Merak (star) GUIDE 8 (*see also* Pointers)
Mercury (messenger of the gods) MERC 22-23; MYTH 8-9
Mercury (planet) ANCIENT 13, 16; ASTER 4, 6-7, 15, 16, 18; EARTH 28; GUIDE 18, 20-21; JUP 18, 28-29; LIFE 16; MARS 28-29; MYTH 8; NEP 6, 7, 28; PILOT 22; PROJ 4, 8, 12, 13; RPS 18-19, 20, 25; SAT 28; SOL 8, 10, 12-13, 18; STAR 22; SUN 28-29; UNIV 6; UR 28; VENUS 20, 28-29; WSP 5, 21
   Atmosphere, level of MERC 10, 28; RPS 19
   Axis of MERC 7, 29
   Core of MERC 8-9
   Length of day MERC 7, 16, 28-29
   Length of year MERC 7, 29
   Magnetic field of MERC 9; VENUS 20
   Orbit of MERC 7, 24-25, 29; PROJ 13
   Orbital tilt MERC 7
   Phases of GUIDE 21; MERC 15, 28
   Rotation of MERC 7, 9, 16, 28; VENUS 20
   (*see also the book* **Mercury: The Quick Planet**)
Mercury (quicksilver) MERC 22-23
"Mercury" dime MERC 22
Mercury missions PILOT 4-5
Meteor Crater, Arizona (*see* Barringer Meteor Crater)
Meteor showers COMET 8-9, 14-15; PROJ 16-17; SOL 20-21; UFO 29 (*see also* names of *individual showers*)
Meteoric iron ASTER 20
Meteorites, meteoroids, and meteors ASTER 19, 20; COMET 6-9, 12-15, 17, 22, 25; DINOS 8-9, 13, 29; EARTH 14; GARB 12, 19, 26, 28-29; JUP 20-21; LIFE 5, 9, 16; MOON 7, 19, 22, 25, 28-29; MYTH 13; PILOT 7; PROJ 8, 16, 17; SOL 20-21; UFO 14, 29; WSP 23 (*see also the book* **Comets and Meteors**)
Meteosat (weather satellite) WSP 9
Methane JUP 8-9; LIFE 4, 17, 18; NEP 14, 17, 20-21, 23; PLUTO 8-9, 16-17, 19, 20-21; SAT 25, 27; SOL 14
Metis (moon of Jupiter) JUP 14-15, 28
Mexico, ancient ANCIENT 8-9 (*see also* Aztecs *and* Mayas)
Mexico, modern EARTH 12-13; RPS 13; WSP 29
Microfilm SF 24, 28
Mid-Atlantic Ridge EARTH 9, 13
Middle East RPS 4

Milky Way (faint band of light) *GUIDE* 12-13, 28-29; *MYTH* 16, 22-23; *PLUTO* 24-25; *PROJ* 20; *STAR* 26-27; *TODAY* 4-5

Milky Way Galaxy *ASTER* 27; *COL* 18-19, 23, 26-27; *DINOS* 22, 23; *GAL* 4-5, 6-7, 8, 10, 12-13, 14-15, 16, 18-19, 20, 22, 24, 25, 26, 28-29; *GUIDE* 12-13, 23, 26-27, 28-29; *LIFE* 8, 17, 20-21, 24, 26; *MYTH* 16, 22-23; *PLUTO* 24-25; *PROJ* 28; *QPB* 14, 22, 26-27, 29; *SF* 19, 22; *SOL* 19, 26, 28; *STAR* 6-7, 11, 17, 26-27; *TODAY* 4-5, 15, 29; *UFO* 27; *UNIV* 8-9, 11, 14-15, 28-29

    Centaurus Arm of *GAL* 7, 18-19, 28-29

    Galactic core of *GAL* 7, 18-19, 24; *QPB* 26

    Orion Arm of *GAL* 6, 7, 18-19, 28-29; *SOL* 28

    Perseus Arm of *GAL* 7, 18-19, 28-29

    Rotation of *GAL* 18; *SOL* 19

    Sagittarius Arm of *GAL* 4-5, 7, 18-19, 28-29

    (*see also the book* **Our Milky Way and Other Galaxies**)

Miller, William (US evangelist) *UFO* 25

Mimas (moon of Saturn) *SAT* 16-17, 20-21, 28-29

Mini-black holes *QPB* 19

Minicomets *COMET* 25

Mining in space *ASTER* 20-21, 23, 24-25; *COL* 10-11, 29; *DINOS* 27; *GARB* 28; *MERC* 17; *MOON* 25-26; *PILOT* 23; *PROJ* 4; *WSP* 25

Minor planets (*see* Asteroids)

Mir (Soviet space station) *PILOT* 15; *SF* 10; *WSP* 17, 29

Mira (star) *GUIDE* 14-15

Miranda (fictional character) *UR* 10

Miranda (moon of Uranus) *LIFE* 16; *UR* 10-11, 20, 24-25, 28-29

"Missing mass" *GAL* 13, 17

MMU *PILOT* 29; *SF* 24-25 (*see also* Space walks)

Molten rock *EARTH* 7, 8-9, 12-13, 16; *UR* 17; *VENUS* 20

Mongolia, ancient *ANCIENT* 7; *MYTH* 23, 25; *RPS* 4

Mongolia, modern *WSP* 29

Montana (US) *DINOS* 4-5

Montezuma (emperor of the Aztecs) *COMET* 20

Moon bases (*see* Space colonies)

"Moon buggy" (*see* Lunar Rover)

Moon, Earth's *ANCIENT* 4, 6, 8-9, 15, 16, 18, 23, 24, 27, 28; *ASTER* 18, 20; *COL* 4-5, 6, 8, 15, 28-29; *COMET* 6-7, 20, 24-25; *DINOS* 27; *EARTH* 6-7, 18-19, 25, 27, 28-29; *GAL* 29; *GARB* 6-7, 12, 21; *GUIDE* 4-5, 6-7, 10, 18, 20, 21, 22, 24, 26; *JUP* 4, 6, 15, 16, 21, 23; *LIFE* 10-11, 14, 16, 24; *MARS* 8-9, 10, 14, 16, 18, 22; *MERC* 4-5, 9, 10, 13, 15, 16, 20-21, 23, 26, 28, 29; *MYTH* 6-7, 8, 12-13, 19; *NEP* 7, 9, 20-21, 23; *PILOT* 8-9, 10, 11, 20, 22, 23, 28-29; *PLUTO* 12-13, 14-15, 19, 21, 28-29; *PROJ* 4, 8, 10-11, 18, 20; *RPS* 7, 9, 16-17, 18-19, 26; *SAT* 16; *SF* 4-5, 7, 8, 12, 13, 24-25, 29; *SOL* 13, 18, 19; *STAR* 22-23; *SUN* 9, 10, 20-21; *TODAY* 5, 14-15, 19, 29; *UFO* 15, 24; *UNIV* 5, 6-7; *UR* 10, 22; *VENUS* 4-5, 7, 9, 21, 28; *WSP* 5, 6, 7, 10-11, 17, 19, 20-21, 24-25, 28

    Atmosphere, lack of *EARTH* 19; *MARS* 22; *MOON* 6, 16, 22, 26

    Core of *MERC* 9; *MOON* 23

    Current geological activity on *GARB* 6; *GUIDE* 26

    Eventual destruction of *STAR* 22-23

    Magnetic field, lack of *MERC* 9; *MOON* 22, 23

    Orbit of *MOON* 8-9, 18, 20-21, 25

    Phases of *ANCIENT* 4, 27; *GUIDE* 6-7, 21; *MERC* 15, 28; *MOON* 8-9; *MYTH* 7; *PROJ* 10-11; *VENUS* 7, 9

    As possible captured planet *MOON* 18

    Rotation of *MOON* 23

    (*see also* Double planet system[s]: Earth-Moon system as, *and the book* **The Earth's Moon**)

Moon goddesses *MYTH* 6-7 (*see also* Artemis, Isis, *and* Selene)

Moon rocks *COMET* 7; *GARB* 12; *MOON* 16, 18, 19; *WSP* 7

    As meteors *COMET* 7

"Moondial" *PROJ* 24

Moonlight *PROJ* 16

"Moonquakes" *GARB* 6

Moons (of other planets) *LIFE* 17; *MOON* 12-13; *PROJ* 8, 12, 20; *SOL* 8-9, 11, 13, 18-19; *SUN* 8

    Captured asteroids as *ASTER* 8-9, 12, 14; *JUP* 15; *MARS* 25; *NEP* 20; *PLUTO* 27; *SAT* 19

    Of Jupiter *ANCIENT* 27; *ASTER* 12, 27; *EARTH* 18, 19; *GUIDE* 23; *JUP* 4-5, 6-7, 12, 14-23, 27, 28-29; *LIFE* 16, 18-19; *MARS* 16; *MERC* 26; *PROJ* 20; *RPS* 24, 25, 27; *SAT* 25; *SOL* 13, 19; *TODAY* 23; *UR* 22 (*see also* Galilean moons of Jupiter, "shepherds," *and names of individual moons*)

    Of Mars *ASTER* 8-9, 11, 22; *GARB* 6, 21; *MARS* 16-17, 18, 22-23, 25, 28-29; *SF* 14; *SOL* 18; *TODAY* 22-23 (*see also* Deimos *and* Phobos)

Of Neptune ASTER 14; LIFE 17, 18; MYTH 11; NEP 8-9, 10-11, 20-21, 22-23, 28-29; PLUTO 10-11, 27; UR 22 (*see also* Nereid, 1989 N1, *and* Triton)
"New" moons NEP 10-11, 29 (*see also* 1989 N1)
Planet-sized LIFE 16-18; MOON 12, 18, 20; NEP 20-21, 22-23; RPS 25, 27; SOL 13, 18-19; UR 22 (*see also* Callisto, Europa, Ganymede, Io, Moon, Earth's, Titan, *and* Triton)
Of Pluto (*see* Charon)
Of Saturn ASTER 14, 27; GUIDE 26; LIFE 16, 18; MARS 16; MERC 26; NEP 21; PILOT 11; PLUTO 12; RPS 27; SAT 13, 14-15, 16, 27, 28-29; SOL 13, 19 (*see also* names of individual moons)
Escaped moons of (*see* Chiron)
Possible lost moon of GUIDE 26
Of Uranus LIFE 16; UR 4, 10-11, 14, 20-25, 26, 28-29
Moonscape MOON 17
Moonwatch MYTH 7
Morelos 1 (satellite) WSP 29
Morning Star ANCIENT 15; GUIDE 21; VENUS 5, 7, 28 (*see also* Evening Star *and* Venus)
Moscow (USSR) MOON 14
Mountains VENUS 22-23 (*see also* names of individual mountains and mountain ranges)
Mountains of the Moon MOON 6
"Mouse" galaxies GAL 23
Movies, science fiction (*see* titles of individual movies)
Multiple stars GAL 10; STAR 9, 11, 17, 21 (*see also* Binary stars, Double-star system[s], *and* Triple stars)
Muslims MOON 20; MYTH 13 (*see also* Arabs, ancient, *and* Black Stone of the Kaaba)

N

Nadel, Aaron (US author) SF 28 (*see also* Amazing Stories *and* "Thought Machine, The")
NASA (National Aeronautics and Space Administration) ASTER 25; COL 7; DINOS 26; GARB 5; PILOT 7, 19, 29; RPS 18-19; WSP 13
Natural gas PLUTO 16-17 (*see also* Methane)
Navigational satellites GARB 29; RPS 14; SF 24, 28; WSP 9
Navajos SUN 22
Nazca (Peru) UFO 9 (*see also* Indians, American)
Nebulas GUIDE 16-17; NEP 24-25; PROJ 20; QPB 4-5; SOL 4-5, 6-7, 8-9, 24-25, 28, 29; STAR 6-7, 8-9, 10-11, 24-25, 28-29; UNIV 7, 24 (*see also* Solar nebula *and* names of individual nebulas)
Nemesis (hypothetical companion star of the Sun) DINOS 20-21, 22; SOL 11
Neptune (Roman god of the sea) MYTH 11; NEP 5
Neptune (planet) ASTER 14; COL 17; EARTH 28-29; GARB 25; GUIDE 23; JUP 28-29; LIFE 17, 18; MARS 29; MERC 29; MOON 12; MYTH 10-11; PLUTO 5, 7, 9, 10-11, 13, 14, 23, 25, 27, 28-29; PROJ 12; RPS 24, 26; SAT 10, 13, 28-29; SOL 10, 14-15, 16-17, 18; SUN 29; UR 6, 16-17, 22, 26, 28-29; VENUS 29; WSP 21
Atmosphere of NEP 4-5, 6-7, 14-15, 16-17, 18
Axis and axial tilt of NEP 19, 20
Clouds of NEP 4-5, 6-7, 14-15, 16-17, 18, 26
Core of NEP 18-19
Equator of NEP 9
Magnetic field of MERC 9; NEP 19
Moons of (*see* Moons: Of Neptune, Nereid, 1989 N1, *and* Triton)
Orbit of NEP 6-7, 15, 28-29; PLUTO 5, 7, 9, 11, 25
Rings of NEP 12-13, 28-29; SAT 13
Rotation of NEP 7, 28
Storms on NEP 4-5, 17 (*see also* Great Dark Spot)
Weather on NEP 15
Wind on NEP 15
(*see also the book* **Neptune: The Farthest Giant**)
Nereid (moon of Neptune) ASTER 14; NEP 8-9, 11, 28-29; PLUTO 27
As a captured asteroid ASTER 14; PLUTO 27
Orbit of NEP 9
Netherlands, The (*see* Holland and the Dutch)
Neutrinos STAR 18; SUN 7; TODAY 14-15
Neutron stars COL 21; GUIDE 25; QPB 8, 10-11, 12-13, 14, 15, 16; STAR 20-21, 28; TODAY 11, 29; UNIV 25-26 (*see also* Pulsars)
New Jersey (US) MARS 7; UFO 17
New Mexico (US) ANCIENT 8; TODAY 13; UFO 23
New York, New York (US) COL 7; RPS 14; SF 8; SUN 27
New York *Sun* (newspaper) LIFE 10-11; UFO 24 (*see also* Great Moon Hoax)
New Zealand UFO 8

Newton, Sir Isaac (British mathematician) COMET 18, 29; RPS 4; SF 8; TODAY 28, 29
NGC 5128 (galaxy) UNIV 25
Niagara Falls COMET 14-15
Nickel-iron alloy ASTER 20 (*see also* Meteoric iron)
Nile River ANCIENT 10-11; GUIDE 16
Nimbus (weather satellites)
    Nimbus 5 RPS 13
    Nimbus 7 RPS 12
1980 S6 (unnamed moon of Saturn) SAT 16-17, 22, 29
1989 N1 (unnamed moon of Neptune) NEP 10-11, 29
Nitrogen EARTH 14; LIFE 4; MARS 12; NEP 23; RPS 27; SAT 25; VENUS 17
Nixon, Richard M. (US president) PILOT 11
Norfolk, Virginia (US) RPS 14-15
Norse, ancient (Vikings) MYTH 12, 16, 23, 25
North Africa RPS 15
North America ASTER 15; COL 15; EARTH 10; MOON 12 (*see also* Americas)
North Pole
    Of Earth GUIDE 5, MOON 9; SUN 18-19
    Of Mars MARS 7
North Star (Polaris, the Pole Star) GUIDE 5, 8-9, 11, 12; MYTH 21, 24-25
    Former pole stars MYTH 25 (*see also* Alderamin, Deneb, Thuban, *and* Vega)
Northern Hemisphere GUIDE 8-9, 28-29; MYTH 20-21
Northern Lights (*see* Aurora Borealis)
"NOVA" (TV series) (*see* "Case of the UFOs, The")
Nova Scotia (Canada) DINOS 16-17
Nova Scotia crater DINOS 16-17
Novas (exploding stars) TODAY 19 (*see also* Exploding stars *and* Supernovas)
Novels and stories, science fiction (*see titles of individual works*)
Nuclear fusion QPB 6-7; SOL 6, 11, 24, 29; STAR 8, 10, 14, 17, 28-29; SUN 6, 11 (*see also* Atomic bomb and atomic energy *and* Hydrogen bomb)
Nuclear war GARB 7, 28-29
Nuclear-powered spacecraft COL 20-21 (*see also* Orion Nuclear-pulse Craft)
Nut (Egyptian goddess of the sky) UNIV 4

O

Oberon (king of the fairies) UR 10

Oberon (moon of Uranus) UR 10-11, 28-29
Observatories, ancient ANCIENT 7, 8-9, 12-13, 16, 28-29; GUIDE 6 (*see also* Caracol, Medicine Wheel, *and* Stonehenge)
Oceans VENUS 15, 17
    Of Earth DINOS 16-17; EARTH 6, 8-9, 10, 15, 20-21, 22; VENUS 17, 25 (*see also* Atlantic Ocean *and* Pacific Ocean)
    Of Venus VENUS 10, 21, 25
Ockels, Wubbo (astronaut) WSP 29
Odyssey, The (Greek epic) SF 5 (*see also* Homer)
Officina Typographica (former constellation) MYTH 29
Oklahoma (US) VENUS 4-5
Olympus Mons (extinct Martian volcano) MARS 10-11, 12-13
Omens (*see* Comets: As omens of disaster)
Oort, Jan COMET 11; DINOS 18; SOL 22
Oort Cloud COMET 10-11, 23; DINOS 18-19, 20-21, 24-25; LIFE 17; PILOT 25; PLUTO 27; SOL 22, 27
Orbital tilt MERC 7; PLUTO 11
Orbits (*see subentries for* Comets, Earth's Moon, *and individual planets*)
Oregon (US) UFO 18-19
Origins of life on Earth (*see* Life, origins of)
Orion (mythological figure) MYTH 16-17, 28
Orion (constellation) GAL 29; GUIDE 16-17; MYTH 5, 16-17, 28; PROJ 16, 20; QPB 5; STAR 11, 12, 14
Orion Nebula GUIDE 16-17; PROJ 20; QPB 4-5; STAR 11
Orion Nuclear-pulse Craft (proposed spacecraft) COL 21
Orionids (meteor shower) PROJ 16
Orrery PLUTO 5
Osumi (satellite) WSP 29
Outer Space Treaty WSP 28
Oxygen COL 8; EARTH 14, 20, 23, 27; LIFE 4, 16-17; MARS 12, 27; PILOT 15, 20; RPS 5, 6; STAR 14, 27; VENUS 15, 17, 25; WSP 4, 15
Ozone layer COL 5; EARTH 22-23; LIFE 16; RPS 12

P

Pacific Ocean ANCIENT 6-7; EARTH 8, 10, 12
Palapa 1 (communications satellite) RSP 10-11
Pallas (goddess) ASTER 11; MYTH 11
Pallas (asteroid) ASTER 11, 29; MYTH 11
Palomar, Mt. TODAY 7
Pan (Greek god) MYTH 28

19

Pandora (moon of Saturn) SAT 16-17, 28-29
Pangaea (supercontinent) EARTH 8
Parícutin (volcano) EARTH 12-13; JUP 22
Paris (France) COMET 16-17, 20-21
Pasiphae (moon of Jupiter) JUP 14-15, 29
Paul, Frank R. (US artist) SF 29 (see also *Amazing Stories*)
Pegasus (constellation) GUIDE 14-15; MYTH 23, 29
Peltier, Leslie (amateur astronomer) TODAY 18-19
Perseids (meteor shower) PROJ 16
Perseus (mythological figure) GAL 28-29; MYTH 25, 28
Perseus (constellation) GAL 28-29; MYTH 25, 28; PROJ 16
Peru UFO 9, 10
Pham Tuan (cosmonaut) WSP 29
Phases
    Of Earth's Moon ANCIENT 4, 27; GUIDE 6-7, 21; MERC 15, 28; MOON 8-9; MYTH 7; PROJ 10-11; VENUS 7, 9
    Of Mercury GUIDE 21; MERC 15, 28
    Of Venus ANCIENT 27; GUIDE 21; MERC 15; PROJ 20; VENUS 6-7, 9, 25
Phobos (moon of Mars) ASTER 8-9; GARB 6, 21; MARS 16-17, 18, 22-23, 25, 28-29; TODAY 22-23
Phobos (Soviet probe) GARB 6; LIFE 16; MARS 17, 18
Phoebe (mythological figure) MYTH 11
Phoebe (moon of Saturn) ASTER 14; MYTH 11; SAT 16-17, 28-29
    As possible captured asteroid ASTER 14
Photo reconnaissance satellites (see Spy satellites)
Photosphere SUN 6, 16
Piazzi, Giuseppe (Italian astronomer) ASTER 7, 11
Piazzia (asteroid) ASTER 11
Pickering, E. C. (US astronomer) VENUS 19
Pioneer deep-space probes GARB 24-25, 26, 28-29; JUP 9, 27; QPB 14; RPS 24-25; WSP 21
    Pioneer 10 GARB 24-25, 26; JUP 9; QPB 14
    Pioneer 11 GARB 24-25; JUP 9; QPB 14
Pioneer Orbiter VENUS 20-21
Pioneer Venus VENUS 13, 21
Pisces (constellation) GUIDE 19
Planet X DINOS 24-25; NEP 25; PLUTO 25, 27; SOL 17; UR 26
Planetariums PROJ 22
Planetary nebulas SOL 29; STAR 24-25 (see also Ring nebulas)
Planetesimals ASTER 28; PROJ 8; UR 9 (see also Chiron)
Planetoids ASTER 28
Planets (see names of and books about individual planets and the book **Our Solar System**)
    In other solar systems (see Solar systems, other)
Plates EARTH 8-9, 10-11, 12-13; VENUS 22
Platypus COL 26
Pleiades (The Seven Sisters) GUIDE 17; PROJ 20
Pluto (cartoon character) MYTH 10; PLUTO 26
Pluto (Roman god of the underworld) MYTH 10-11; PLUTO 6-7, 13
Pluto (planet) ASTER 14-15; COL 17, 18; COMET 10-11; EARTH 28-29; GUIDE 23; JUP 29; LIFE 17; MARS 29; MERC 7, 26, 28-29; MOON 13; MYTH 10-11; NEP 7, 15, 23, 24-25, 26-27, 28-29; PILOT 25; PROJ 12, 13; SAT 28-29; SOL 10, 16-17, 18, 21, 22; SUN 29; TODAY 23; UNIV 29; UR 26, 28-29; VENUS 11, 29; WSP 21
    As asteroid ASTER 14, 15
    Atmosphere of MOON 13; PLUTO 14, 17, 19, 20-21, 28
    Discovery of MYTH 10-11; PLUTO 7, 14; TODAY 23; UR 26
    Equator of PLUTO 17
    As escaped moon of Neptune PLUTO 10-11
    Orbit of MERC 7; NEP 15, 28; PLUTO 9, 10-11, 18, 28-29; SOL 16; VENUS 11
    Orbital tilt MERC 7; PLUTO 11
    Poles of PLUTO 17
    Rotation of PLUTO 17, 19, 28
    (see also the book **Pluto: A Double Planet?**)
Pluto-Charon system MOON 13; PLUTO 12-13, 18-19, 23 (see also Double Planet[s]: Pluto-Charon system as)
Poe, Edgar Allan (US author) WSP 25
Pointers GUIDE 8-9 (see also Big Dipper)
Poland COMET 28; WSP 29
Polar ice caps
    Of Earth EARTH 18-19 (see also Antarctic Circle, Antarctica, and Arctic Circle)
    Of Mars EARTH 18-19; MARS 4, 6-7, 20, 24, 26-27; RPS 22
    Of Triton NEP 23
Polaris (star) (see North Star)
Pole Star (see North Star)
Pole stars, former MYTH 25 (see also Alderamin, Deneb, Thuban, and Vega)
Pole-ring galaxies SOL 27

Poles (*see subentries for individual planets and* Axis *subentries for individual planets*)
Pollution   EARTH   22-23, 24; WSP   23 (*see also the book* **Space Garbage**)
Pollux (star)   STAR   13
Polynesians, ancient   ANCIENT   6-7
Pope, Alexander (British poet)   UR   10
Poseidon (Greek god of the sea)   NEP   9 (*see also* Neptune [Roman god of the sea])
Potatoes   MARS   16
Praying mantis   LIFE   22
Prisms   PROJ   14-15; TODAY   28, 29
Probes   ASTER   8, 22; COMET   19, 26-27, 29; EARTH   25; GARB   21, 24-27, 28-29; JUP   9, 15, 16, 21, 23, 25, 26-27; LIFE   14, 16-17; MARS   9, 18-19; MERC   14-15, 26-27; MOON   14-15, 28; NEP   12, 26-27; PILOT   8-9; PLUTO   22-23, 25; PROJ   4, 6, 28; RPS   16-27; SAT   10, 12, 27; SF   18-19; SUN   26; TODAY   23, 28; UR   16, 25, 26; VENUS   14-15, 17, 18-19, 20-21, 22-23, 28-29; WSP   5, 20-21 (*see also names of individual probes and the book* **Rockets, Probes, and Satellites**)
Procyon (star)   STAR   13
Progress (Soviet spacecraft)   COL   13
Prometheus (mythological figure)   MYTH   11
Prometheus (moon of Saturn)   MYTH   11; SAT   16-17, 28-29
Prominences   SUN   6, 16, 28
Protoplanets   SOL   8-9
Protostars   STAR   6-8, 28-29
   The Sun as a protostar   SOL   5, 24-25, 29 (*see also* Gas clouds, stellar)
Prunariu, Dumitru (cosmonaut)   WSP   29
Psychokinesis   SF   9
Ptolemy (Greek-Egyptian astronomer)   ANCIENT   16-17, 20; VENUS   7
Puerto Rico   SF   22
Pulsars   QPB   12-13, 14, 16; TODAY   11, 29; UNIV   3, 13 (*see also* Neutron stars *and* Variable stars)
Pulsating stars (*see* Pulsars)
Pyramid, Great   UFO   8 (*see also* Egyptians, ancient)
Pythagoras (Greek philosopher)   ANCIENT   14-15

Q

Quadrantids (meteor shower)   PROJ   16
Quasars   GAL   25; QPB   22-26, 28-29; TODAY   11, 15, 25, 29; UNIV   3, 12-13, 15, 22-23, 29
"Quick Planet" (*see* Mercury [planet])
Quicksilver (mercury)   MERC   22-23

R

Ra (Egyptian Sun god)   MYTH   4-5; SUN   10-11
Ra-Shalom (asteroid)   ASTER   28
Radar   VENUS   20-21, 23
Radar tracking   GARB   10-11; UFO   11
Radiation   GAL   24-25, 26; GARB   12, 16-17; MERC   16-17; QPB   16, 21; STAR   18; TODAY   4-5, 28-29; VENUS   13; WSP   22-23 (*see also* Cosmic rays, Gamma rays, Infrared radiation, Solar radiation, Ultraviolet light, *and* X-rays)
Radio astronomy   TODAY   29; VENUS   13, 19
Radio beams (*see* Radio waves)
Radio mapping   RPS   20
Radio signals or messages   LIFE   8, 17; QPB   13; SF   22-23; TODAY   11 (*see also* Arecibo Message)
Radio telescopes   GARB   10-11; LIFE   8, 17; MOON   26; QPB   17; SF   22; TODAY   11, 12-13, 28; UNIV   12; VENUS   13; WSP   24
Radio waves   GAL   24; GARB   11; LIFE   8; MERC   14-15; PROJ   14; QPB   12-13, 16, 21, 22, 26; RPS   11-12, 20; TODAY   10-11, 28, 29; UNIV   12, 21; UR   18; VENUS   13, 18-19, 21
Rahu (dragon)   MYTH   13
Rain forests   EARTH   22-23
Rainbows   PROJ   14; QPB   24; SUN   23; TODAY   28
Rainier, Mt.   MARS   13
Ralph 124C 41+ (novel)   SF   28-29 (*see also* Gernsback, Hugo)
"Rattail" galaxies   GAL   26
Rays (lunar features)   JUP   19; PROJ   8
Red dwarfs (stars)   GARB   25; STAR   28-29
Red giant(s) (stars)   GAL   8-9; GUIDE   12, 16; QPB   8; SOL   24, 25, 29; STAR   12, 13, 14-15, 17, 20-21, 22-23, 24, 25, 29; SUN   21
   The Sun as   GAL   8; SOL   24-25, 29; STAR   22-23, 24, 29
"Red Planet" (*see* Mars [planet])
Red shift   QPB   24-25; TODAY   28, 29; UNIV   10-11, 12 (*see also* Violet shift)
Red supergiants (stars)   GARB   26; STAR   14, 28 (*see also* Supergiants)
Regulus (star)   GUIDE   10-11
Relativity, General Theory of   TODAY   25 (*see* Einstein, Albert)
Remek, Vladimir (cosmonaut)   WSP   29
Reptiles   DINOS   5, 25
Resnik, Judith (astronaut)   PILOT   17; WSP   13
Rhea (moon of Saturn)   SAT   16-17, 20, 28-29

Richmond (Virginia) UFO 28
Ride, Sally (astronaut) PILOT 17; WSP 13
Rigel (star) GUIDE 16-17; STAR 12
Ring Nebula (in Lyra) STAR 24
Ring nebulas STAR 24-25
"Ring of Fire" EARTH 12 (see also Volcanoes and volcanic activity)
Rings
    Of Earth MOON 18
    Of Jupiter JUP 12-13, 14-15, 29; NEP 12; RPS 25; SAT 13; SOL 21
    Of Neptune NEP 12-13, 28-29; SAT 13
    Of Saturn GARB 24-25; GUIDE 23; JUP 12; NEP 12; RPS 24-25; SAT 4-5, 10-11, 12-13, 14-15, 16-17, 21, 22-23; UR 13, 21
    Of Uranus NEP 12; SAT 13; UR 4, 11, 12-13, 14, 20-21, 23, 25, 26
Ritter, Johann Wilhelm (German scientist) TODAY 29
Robots and robotics COL 12-13; GARB 20; SF 7, 25, 29
Rockefellia (asteroid) ASTER 11
Rocket belts SF 24-25 (see also MMU)
Rockets EARTH 25, 26; MOON 14, 16; NEP 6; PILOT 5, 13, 27; PROJ 4; SF 7, 8, 24; UFO 15; WSP 4-5, 6-7, 12, 13, 14, 15, 19, 20, 25, 28 (see also Saturn V, V-2, and the book **Rockets, Probes, and Satellites**)
Rocky Mountains EARTH 10
Rocky planets NEP 7; PROJ 12; SOL 8, 12-13, 18 (see also planet entries Earth, Mars, Mercury, Pluto, and Venus)
Roman Empire COMET 21, 28; WSP 25
Romans, ancient COMET 21, 28; MERC 22; MYTH 8; NEP 5; SF 4; VENUS 5, 7
Rosette Nebula STAR 8-9
Ross 248 (red dwarf star) GARB 25
Roswell (New Mexico) UFO 23
Roswell Incident, The (book) UFO 23
Rover, Lunar (see Lunar Rover)
Rover, Martian MARS 19, 24
Royal Astronomical Society UR 5
R.U.R. (play) SF 29 (see also Capek, Karel)
Russia (see Soviet Union)
Russian (language) MARS 20

S

Sagittarius (constellation) GAL 7, 29; GUIDE 12, 13, 19
St. Helens, Mt. (volcano) EARTH 12; MARS 13

Salisbury, England PROJ 24
Salyut (Soviet space station) GARB 9; PILOT 15; WSP 17, 28-29
Samurai Warriors (Japanese constellation) MYTH 16 (see also Orion [constellation])
San Andreas Fault EARTH 11
San Marco I (satellite) WSP 28
Satan VENUS 7
Satellites, artificial GARB 4-5, 11, 12, 14-15, 16-17, 19, 23, 28-29; GUIDE 21, 24; PILOT 5, 28; SF 17, 24, 27-28; SUN 24, 27; UNIV 8-9; WSP 8-9, 10-11, 12, 13, 16, 28-29 (see also Communications satellites, Navigational satellites, Spy satellites, Weather satellites, *names of individual satellites*, and the book **Rockets, Probes, and Satellites**)
Satellites, natural (see Moon, Earth's and Moons)
Saturn (Roman god of agriculture) MYTH 8, 11; SAT 5; UR 11
Saturn (planet) ANCIENT 7, 13, 16; ASTER 6-7, 14, 27, 28; COL 17; COMET 22-23; EARTH 28-29; GARB 24-25; GUIDE 18, 23, 26; JUP 9, 12, 29; LIFE 16, 18; MARS 16, 29; MERC 26, 29; MOON 12; MYTH 8, 11; NEP 7, 12, 15, 19, 21, 24-25, 28-29; PILOT 11, 25; PLUTO 12, 23, 28-29; PROJ 12; RPS 24-25, 27; SOL 10, 13, 14-15, 18-19; SUN 29; UNIV 6; UR 5, 6, 11, 13, 14, 16, 17, 18, 21, 22, 28-29; VENUS 29; WSP 21
    Atmosphere of SAT 10-11, 12-13, 27; UR 18
    Axial tilt of SAT 28
    Belts of UR 18
    Clouds of SAT 10; UR 18
    Core of SAT 8, 10
    Equator of SAT 7, 12
    Magnetic field of MERC 9; NEP 19
    Orbit of SAT 28-29
    Rings of GARB 24-25; GUIDE 23; JUP 12; NEP 12; RPS 24-25; SAT 4-5, 10-11, 12-13, 14-15, 16-17, 21, 22-23; UR 13, 21
    Rotation of SAT 7, 28
    Wind on NEP 15
    (see also the book **Saturn: The Ringed Beauty**)
Saturn V (rocket) UFO 15; WSP 5
Saturn system, the SAT 16-17 (see also Moons: Of Saturn and Rings: Of Saturn)
Saudi Arabia WSP 29
Savitskaya, Svetlana (cosmonaut) PILOT 6-7
Saxons (ancient English) COMET 21
SBTS (see Brasilsat 1)
Scandinavia EARTH 21

Scarps (cliffs) *UR* 24-25
Schwabe, S. H. (German astronomer) *TODAY* 23
Schweickart, Russell L. (astronaut) *MOON* 16
Science fiction *ASTER* 26-27; *LIFE* 12-13;
  PROJ 4; *UFO* 7, 17, 20; *VENUS* 10 (*see
  also the book* **Science Fiction, Science Fact**)
Scorpio (constellation) *GUIDE* 12-13, 19; *MYTH*
  28; *STAR* 25
Scorpius (*see* Scorpio)
Scott, David R. (astronaut) *MOON* 16-17
Sea of Tranquility (lunar feature) *MOON* 17
Sears Tower (Chicago) *EARTH* 29
"Seas," lunar *MOON* 6, 14, 17, 19, 29; *PROJ* 8
  (*see also* Mare Moscoviense, Mare Orientale,
  and Sea of Tranquility)
Second-generation stars *STAR* 27
Selene (Greek Moon goddess) *MYTH* 6-7
SETI (Search for Extraterrestrial Intelligence) *LIFE*
  8 (*see also* Alien Life, Arecibo Message, *and*
  "Little Green Men")
Settlements in space (*see* Space colonies)
Settlements on other planets (*see* Space colonies)
Seven Boys Transformed into Geese (American
  Indian constellation) *MYTH* 29 (*see also*
  Big Dipper)
Seven Sisters (*see* Pleiades)
Sextant *MYTH* 21
Shakespeare, William (British playwright) *UR* 10
Shamash (Babylonian Sun god) *MYTH* 5
Sharks *LIFE* 22
Shelley, Mary (British author) (*see Frankenstein*)
Shepard, Alan (astronaut) *PILOT* 4-5; *WSP* 28
"Shepherds" (moons) *JUP* 12; *SAT* 15, 22-23
"Shooting stars" *COMET* 6-7; *MYTH* 13 (*see
  also* Meteorites, Meteoroids, *and* Meteors)
"Siamese Twin" galaxies *GAL* 17
Siberia *COMET* 25; *DINOS* 14-15; *EARTH*
  21; *MARS* 7 (*see also* Tunguska Incident)
Silicon *STAR* 27
Sinope (moon of Jupiter) *JUP* 14-15, 29
Sirius (star) *ANCIENT* 10-11; *GUIDE* 16-17;
  MYTH 5; *STAR* 12-13
Skidi Pawnee Indians *MYTH* 23, 29
Skylab (US space station) *COL* 7; *GARB* 8, 11,
  16-19, 29; *PILOT* 21; *SF* 10-11; *SUN* 8,
  17, 24; *WSP* 17
"Skylark of Space, The" (story) *SF* 7
Skyscrapers *SF* 28-29
Slipher, Vesto Melvin (US astronomer)
  TODAY 29
Small Magellanic Cloud (galaxy) *GAL* 12-13;
  GUIDE 28-29
Smog *SAT* 24-25
Snakes *DINOS* 5

Socorro (New Mexico) *TODAY* 13
Soil, lunar *MOON* 16-17
Solar eclipses (*see* Eclipses: Solar)
Solar energy *WSP* 16-17, 22
Solar eruptions *SUN* 8 (*see also* Prominences)
Solar flares *GARB* 12, 16-17, 19, 28-29; *STAR*
  25; *SUN* 14, 15, 16, 17, 28
Solar Maximum (satellite) *GARB* 12
Solar nebula *QPB* 4-5; *SOL* 5, 6-7, 8-9, 24-25,
  28-29; *STAR* 6-7, 8-9, 10-11, 28-29; *UNIV*
  7, 24 (*see also* Nebulas)
Solar panels *WSP* 16, 22, 25
Solar radiation *MARS* 22; *MERC* 16-17;
  MOON 22, 25; *SUN* 24; *WSP* 23
Solar spectrum *SUN* 23
Solar system *ASTER* 6, 14, 22, 27, 28; *COL* 11,
  16-18, 23, 25, 28-29; *COMET* 10-11, 12, 27,
  29; *DINOS* 12, 19, 20-21, 22, 24-25, 26;
  EARTH 4-5, 19, 27, 28-29; *GAL* 6, 19;
  GARB 24-25, 26, 29; *GUIDE* 23; *JUP* 7, 9,
  16, 18, 28-29; *LIFE* 8, 9, 14, 16-17, 18, 21;
  MARS 10, 11, 28-29; *MERC* 8, 28-29;
  MOON 12, 26; *NEP* 7, 9, 11, 20-21, 24-25,
  26-27, 28-29; *PILOT* 25; *PLUTO* 4, 5, 6,
  11, 12, 13, 23, 25, 27, 28-29; *PROJ* 4, 8, 12-
  13; *QPB* 19, 29; *RPS* 19, 20, 24-25, 27;
  SAT 7, 20, 28-29; *SF* 8, 14, 22-23; *SUN* 4,
  8, 28-29; *TODAY* 11; *UFO* 18; *UNIV* 6-
  7, 14-15, 28-29; *UR* 28-29; *VENUS* 16-17,
  28-29; *WSP* 19, 21
  Origins of *DINOS* 19; *EARTH* 4-5; *SOL*
    5-9; *SUN* 4-5; *UNIV* 7; *UR* 9 (*see also*
    Solar nebula)
  (*see also the book* **Our Solar System**)
Solar systems, other *COL* 18-19, 25, 27; *EARTH*
  20, 27; *GAL* 10, 17, 22; *LIFE* 17, 21, 24-25;
  PILOT 24-25; *PROJ* 4; *SOL* 26-27; *STAR*
  9, 14-15, 29; *TODAY* 10-11; *UFO* 18; *WSP*
  23 (*see also* Beta Pictoris)
Solar telescopes (*see* McMath Solar Telescope)
Solar transits *MERC* 18-19; *VENUS* 8-9
Solar wind *EARTH* 16-17; *GARB* 12, 21, 28-
  29; *NEP* 27; *SOL* 8-9, 22, 29; *SUN* 6, 18-
  19, 25
Solstice, summer *GUIDE* 6
Sombrero Galaxy *DINOS* 22-23; *GAL* 20-21
"Sounds of Earth, The" (record) *PROJ* 28;
  RPS 25
South Africa *DINOS* 25
South America *COL* 15; *MOON* 12; *MYTH*
  22-23 (*see also* Americas)
South Dakota (US) *PROJ* 14
South Korea *TODAY* 5
South Pole (Earth's) *GUIDE* 8; *SUN* 18-19

Southern Cross (constellation) GUIDE 8, 28-29; STAR 12
Southern Hemisphere GAL 12, 20; GUIDE 28-29; MOON 13; MYTH 20-21, 29; STAR 18
Southern Lights (see Aurora Australis)
Soviet Union (USSR) COL 7, 13; COMET 27; GARB 6-7, 9, 16, 17, 21; LIFE 16; MARS 17, 18-19, 20; MOON 14-15, 16, 19, 26; PILOT 4-5, 6-7, 8-9, 13, 15, 16-17, 20, 22, 26, 27; RPS 4, 8-9, 16, 21, 28-29; SF 27; SUN 24; TODAY 7; VENUS 14-15, 22, 28; WSP 5, 9, 13, 17, 28, 29
Soyuz missions PILOT 16-17, 26; WSP 29 (see also Apollo-Soyuz Test Project)
Space Camps PILOT 28-29
Space colonies ASTER 26-27; DINOS 26-27; EARTH 25-27; GARB 27, 29; LIFE 17, 26-27; MARS 22-23, 27; MOON 22-27; PILOT 23, 25; PROJ 4; RPS 26, 27; SF 22; WSP 19, 21, 24, 25 (see also the book **Colonizing the Planets and Stars**)
Space junk UFO 15 (see also the book **Space Garbage**)
Space Mirror (proposed satellite) COL 7; DINOS 26; SUN 27
Space shuttles GARB 12-13, 23; PILOT 12-13, 18, 20, 26, 28-29; RPS 10-11, 15, 27; SF 17, 24-25; WSP 12-13, 14, 15, 28-29 (see also Atlantis, Buran, Challenger, Columbia, Discovery, and Hermes)
Space stations COL 6-7; EARTH 25, 26; GARB 8, 11, 16-19, 29; PILOT 15, 21; PROJ 4; RPS 10, 26-27, 28-29; SF 10-11, 16-17; SUN 8, 17, 24; WSP 16-17, 18-19, 28-29 (see also Freedom, Mir, Salyut, and Skylab)
Space suits MERC 6-7; PILOT 7, 28; SF 24-25, 29
Space Surveillance Center (US) GARB 11
Space telescopes PROJ 18; SF 17; WSP 9, 16, 22-23 (see also Hubble Space Telescope)
Space walks PILOT 6-7, 29
Spaceplanes WSP 14-15
Spain WSP 29
Spaniards ANCIENT 9
Spationautes (French astronauts) WSP 28 (see also Astronauts and Cosmonauts)
Spectral lines SUN 23; TODAY 29
Spectrogram SUN 23
Spectroheliograph SUN 22
Spectroscope SUN 22; TODAY 28, 29
Spectrum PROJ 14-15; QPB 24; SUN 22-23; TODAY 4-5, 28-29
Speed of light COL 18; GAL 29; GUIDE 23; SF 20-21, 22; TODAY 14-15 (see also Light-years)
Spica (star) GUIDE 10-11
Spiral arms of galaxies
    Of Andromeda Galaxy GAL 20
    Of Milky Way Galaxy GAL 4-5, 6-7, 18-19, 28-29; SOL 28 (see also subentries under Milky Way Galaxy)
Spiral galaxies GAL 6-7, 14-15, 18-19, 20-21, 22, 25; GUIDE 26; QPB 23, 28-29; SOL 23; UNIV 8-9
Spur (of galaxy) GAL 19 (see also Spiral arms of galaxies)
Sputnik 1 (Earth's first artificial satellite) GARB 25; RPS 8-9; WSP 5, 28
Sputnik 2 RPS 9
Spy satellites RPS 14-15, 28-29; WSP 9
Spyglass PROJ 18 (see also Telescopes)
Square of Pegasus (constellation) GUIDE 14-15 (see also Pegasus)
SS 433 (star) QPB 16-17
Star colonists (see Space colonies)
Star maps, ancient ANCIENT 7, 17; GUIDE 11, 19; MYTH 23
Star maps and star wheels GUIDE 8-9, 11, 13, 15, 17, 29; PROJ 16, 20, 22 (see also Skidi Pawnee Indians and Zodiac of Dendera)
Star of Bethlehem COMET 28
    As possible comet COMET 28
Star Probe (probe) SUN 26
"Star Trek" (TV series) LIFE 12; SF 11
Star Wars (movie) LIFE 12; SF 26
"Star Wars" (satellite system) (see Strategic Defense Initiative)
"Stardial" PROJ 24
"Star-Spangled Banner, The" (song) RPS 5
Stickney, Angelina (Mrs. Asaph Hall) MARS 16
Stickney (crater on Phobos) MARS 16
Stonehenge ANCIENT 28; GUIDE 6; PROJ 24
Stories, science fiction (see individual titles of works)
Storms of Jupiter (see Great Red Spot and Jupiter: Storms on)
Storms of Neptune (see Great Dark Spot and Neptune: Storms on)
Storms of the Sun (see Sunspots)
Strategic Defense Initiative (satellite system) GARB 28-29
Stratos PROJ 6
Styx (mythological river) MYTH 10
Submarines SF 4, 28-29
Sulfur JUP 22-23, 24-25; LIFE 4; QPB 12
Sulfuric acid LIFE 16; RPS 20; VENUS 15
Sumerians MYTH 16, 23

Summer solstice (*see* Solstice, summer)
Summer Triangle (constellation) GUIDE 12
Sun ANCIENT 6, 9, 11, 15, 16, 18, 22-23, 24, 28; ASTER 4, 5, 6-7, 9, 12-13, 16-17, 18, 22, 27, 28; COL 7, 11, 13, 17, 19; COMET 5, 10-11, 12-13, 17, 18, 22-23, 27; DINOS 10-11, 18-19, 20-21, 22-23, 24, 29; EARTH 4-5, 15, 16, 23, 27, 28-29; GAL 6-7, 8-9, 10, 18, 25; GARB 12, 16-17, 19, 25; GUIDE 4-5, 6, 10, 12, 16, 18-19, 21, 23, 25; JUP 4, 6, 25, 27, 28-29; LIFE 13, 16, 21; MARS 4-5, 28-29; MERC 5, 7, 8, 9, 15, 16-17, 18-19, 21, 23, 25, 28-29; MOON 5, 8, 10-11, 25; MYTH 4-5, 6, 8, 11, 12-13, 18-19, 21; NEP 6-7, 9, 14-15, 23, 27, 28-29; PILOT 7; PLUTO 4-5, 6, 9, 10, 11, 20-21, 27, 28-29; PROJ 4, 10, 11, 12, 13, 14, 18, 24, 26; QPB 4-5, 6-7, 10-11, 15, 21, 27, 29; RPS 12, 17, 18, 20, 22, 24, 27; SAT 5, 7, 14, 23, 28-29; SOL 4-5, 6-7, 8-9, 10-11, 12-13, 14, 16, 19, 20, 21, 22, 24-25, 26, 28, 29; STAR 5, 8, 10-11, 12, 14-15, 18, 20-21, 22-23, 24, 27, 28-29; TODAY 5, 7, 15, 16, 19, 21, 23, 25, 28, 29; UFO 6, 27; UNIV 4-5, 6-7, 12, 14-15, 22-23, 24, 26, 28-29; UR 5, 6, 8, 16-17, 26; VENUS 5, 7, 9, 10, 11, 17, 27, 28-29; WSP 16, 26-27
  Atmosphere of (*see* Corona)
  As a black dwarf SOL 29; STAR 29
  Eclipses of (*see* Eclipses: Solar)
  Galactic orbit of GAL 18; SOL 19
  As a god MYTH 5; SUN 10-11
  Magnetic fields of SUN 14, 16
  In possible double-star system DINOS 20-21, 22; SOL 11 (*see also* Nemesis)
  As a red giant GAL 8; SOL 24-25, 29; STAR 22-23, 24, 29
  As a white dwarf GAL 8-9; QPB 10; SOL 24-25, 29; STAR 20, 24, 29
  (*see also the book* **The Sun**)
Sun gods MYTH 4-5; SUN 10-11, UNIV 4 (*see also* Apollo, Aton, Helios, Ra, *and* Shamash)
Sundials ANCIENT 11; PROJ 24-25
"Sun-grazers" (comets) COMET 22-23
Sunlight ANCIENT 15, 18; COL 7; DINOS 10-11, 26-27, 29; GUIDE 6; LIFE 4, 9; MARS 6, 27; MERC 6-7, 20-21; MOON 5, 8-9, 10-11, 22, 27; MYTH 5, 11, 12; NEP 7, 20-21, 23; PLUTO 6, 20-21; PROJ 14-15, 24; RPS 9, 17; SAT 14-15, 27; SOL 11; STAR 10-11; SUN 4, 6, 10-11, 17, 22-23, 27; TODAY 9, 28, 29; UNIV 5, 22; UR 16, 18, 21, 28; VENUS 10, 17, 26-27; WSP 16

Sunspot cycle SUN 17, 26-27; TODAY 23
Sunspots PROJ 26-27; STAR 25; SUN 6, 13, 14-15, 16-17, 22, 26-27, 28
Supergiants (stars) COL 21; GARB 26; STAR 14, 28 (*see also* Red supergiants)
Supernova 1987A GAL 9; STAR 18-19; TODAY 15
Supernovas ANCIENT 17, 21, 23, 24-25; GAL 8-9; GARB 21; QPB 8-9, 12; SOL 4; STAR 16-17, 18-19, 20, 24, 27, 28; SUN 4-5; TODAY 15; UNIV 7, 24-25, 26
Surveillance satellites (*see* Spy satellites)
Surveyor (probes) MOON 15; PILOT 9
  Surveyor 3 PILOT 9
  Surveyor 6 MOON 15
Swamps SF 14; VENUS 10-11
Sweden WSP 29
Swift, Jonathan (British author) SF 14 (*see also* Gulliver's Travels)
Switzerland UFO 6
Syria WSP 29

T

Tape recorders SF 24, 29
Tar LIFE 6; SAT 24, 27
Taurus (constellation) GUIDE 16-17, 19; MYTH 9, 17, 18; UR 13
Telecommunications (*see* Communications satellites)
Telescopes ANCIENT 26-27, 28-29; ASTER 5, 11, 16, 22; EARTH 19, 25; GAL 5, 15; GUIDE 12, 14, 21, 22-23, 24-25, 26; JUP 4-5, 6-7, 10, 12; MARS 6, 11; MERC 15, 19; MOON 6-7, 10, 26; NEP 12, 28; PROJ 18-19, 20, 26; SAT 4-5, 7, 16; STAR 17, 24; SUN 22; TODAY 6-7, 8-9, 13, 16-17, 20-21, 22-23, 28, 29; UFO 17; UNIV 8-9, 24; UR 5, 14; VENUS 7, 9; WSP 22, 23
  Reflecting GUIDE 24-25
  Refracting GUIDE 24
  (*see also* Hubble Space Telescope *and* Radio telescopes)
Telescopium (constellation) GUIDE 29
Telescopium Herschel II (former constellation) MYTH 29
Telesto (moon of Saturn) SAT 22, 28-29
Television MOON 17; PROJ 4; SF 12-13, 19, 24, 29; WSP 10, 11
Tereshkova, Valentina (cosmonaut) PILOT 17; WSP 5
Terraforming ASTER 22; EARTH 27; MARS 26-27; MOON 27; VENUS 25

25

Test-tube babies SF 29
Tethys (moon of Saturn) SAT 13, 16-17, 20-21, 22, 28-29
Texas (US) MOON 7
Thales (Greek astronomer) ANCIENT 15
Tharsis ridge (Martian mountain range) MARS 12
Thebe (moon of Jupiter) JUP 14-15, 28-29
34 Tauri UR 13 (see also Uranus)
"Thought Machine, The" (short story) SF 28 (see also *Amazing Stories*, Computers, and Nadel, Aaron)
Thuban (former pole star) MYTH 25
Thunderstorms EARTH 15
Tiamat (Babylonian goddess) MYTH 28
Tiamat (Babylonian constellation) MYTH 28
Tides MOON 4, 8, 25
Time capsules PROJ 28-29
Time Machine, The (novel) SF 29 (see also Wells, H. G.)
Time travel SF 14, 20-21, 28
Tirawahat (American Indian god) MYTH 29
Tiros 8 (weather satellite) RPS 12
Titan (moon of Saturn) LIFE 16, 18; MERC 26; NEP 21; PILOT 11; PLUTO 12; RPS 27; SAT 16-17, 24-27, 28-29; SOL 13, 19; UR 22
    Atmosphere of LIFE 16, 18; NEP 21; PILOT 11; RPS 27; SAT 24-25, 26-27
Titania (queen of the fairies) UR 10
Titania (moon of Uranus) UR 10-11, 22-23, 28-29
Titius, Johann Daniel (German mathematician) UR 6 (see also Bode's Law)
Titov, Gherman (cosmonaut) PILOT 5
"Toadstool" galaxies GAL 27
Tombaugh, Clyde (US astronomer) NEP 24-25; PLUTO 7; TODAY 23
Tornadoes EARTH 15
Torricelli, Evangelista (Italian scientist) WSP 19
Tower of Babel ANCIENT 12-13 (see also Bible)
Transits (see Solar transits)
Trent, Mr. and Mrs. Paul (UFO spotters) UFO 18-19
Trindade Island (Brazil) UFO 19
Trip to the Moon, A (movie) SF 4-5
Triple stars GAL 10; SOL 11 (see also Mizar and Multiple stars)
Tritium (see Hydrogen)
Triton (Roman sea god, son of Neptune) MYTH 11; NEP 9
Triton (moon of Neptune) LIFE 17, 18; MYTH 11; NEP 8-9, 20-21, 22-23, 28-29; UR 22
    As asteroid NEP 20
    Atmosphere of LIFE 17, 18; NEP 21, 22-23
    As giant comet NEP 20
    Orbit and orbital tilt of NEP 9
    Polar cap of NEP 23
Trojan War ASTER 12
Trojans (asteroids) ASTER 6-7, 12-13; SOL 21
Tsiolkovsky, Konstantin (Russian space theorist) EARTH 25; RPS 6, SF 16; WSP 5
Tuan, Pham (see Pham Tuan)
Tube worms LIFE 19
Tucana (constellation) GUIDE 29
Tucson (Arizona) COMET 4-5
Tulsa (Oklahoma) VENUS 4-5
Tung-Fang-Hung (satellite) WSP 29
Tunguska Incident (Siberia, USSR) COMET 25; DINOS 14-15
Tunguska River, Siberia (USSR) DINOS 15 (see also Siberia)
Turks COMET 21, 29
Turtles DINOS 5
Twenty Thousand Leagues Under the Sea (novel) SF 28 (see also Verne, Jules)
Twin planets, Earth and Venus as LIFE 13, 16; VENUS 10-11, 17, 25, 27, 28
II Zwicky 23 (galaxy) UNIV 24
Tyrannosaurus (dinosaur) DINOS 4-5, 28

U

UFOs (Unidentified Flying Objects) GARB 9
    Alleged kidnappings by UFO 20-21, 24-25, 29
    Government investigations of UFO 22-23
    Hoaxes about UFO 16-17, 20, 22-23, 24-25
    (see also the book **Unidentified Flying Objects**)
Uhuru (satellite) RPS 29
Ultrasaurus (dinosaur) DINOS 7
Ultraviolet light EARTH 23; MERC 16; PROJ 14, 18; RPS 12; SUN 25; TODAY 4-5, 28, 29; VENUS 13
Ultraviolet radiation, rays, and waves (see Ultraviolet light)
Ulysses (probe) RPS 16
Umbriel (fairy) UR 10
Umbriel (moon of Uranus) SF 15; UR 10-11, 22, 28-29
Unidentified Flying Objects (see UFOs)
Union of Soviet Socialist Republics (USSR) (see Soviet Union)
United Kingdom (see Britain and the British)
United Nations SF 12; WSP 28
United States (US) COL 9, 15; EARTH 21; GARB 6-7, 8, 12-13, 18-19; MARS 18-20; MOON 14-15, 16-17, 19, 26; PILOT 5, 7, 9,

12-13, 14-15, 16-17, 22, 26; *QPB* 22; *RPS* 4, 8, 15, 16, 28-29; *SF* 13, 16, 27; *SUN* 24; *TODAY* 29; *UFO* 8-9, 22-23; *VENUS* 15, 22, 23, 28; *WSP* 5, 7, 9, 13, 19, 21, 28, 29
United States Air Force *UFO* 23
University of Arizona *QPB* 25
Uranus (Greek god of the sky) *MYTH* 11; *UR* 3, 11
Uranus (planet) *ASTER* 14, 28; *COL* 17; *EARTH* 28-29; *GARB* 25; *GUIDE* 23; *JUP* 29; *LIFE* 16; *MARS* 28-29; *MERC* 29; *MYTH* 11, 29; *NEP* 5, 7, 12, 15, 19, 24-25, 28-29; *PILOT* 25; *PLUTO* 5, 7, 14, 25, 28-29; *PROJ* 12; *RPS* 24, 26; *SAT* 10, 13, 28-29; *SF* 15; *SOL* 10, 14-15, 18; *SUN* 29; *UNIV* 8; *VENUS* 28-29; *WSP* 21
   Atmosphere of *SOL* 14; *UR* 13, 15, 17, 18-19
   Axis and axial tilt of *NEP* 19; *UR* 8-9, 10-11, 15, 18, 28
   Clouds of *UR* 18-19
   Core of *UR* 17
   Magnetic field of *MERC* 9; *NEP* 19
   Moons of (*see* Moons: *Of Uranus and names of individual moons*)
   Orbit of *NEP* 5; *PLUTO* 5, 7, 25; *UR* 26, 28-29
   Rings of *NEP* 12; *SAT* 13; *UR* 4, 11, 12-13, 14, 20-21, 23, 25, 26
   Rotation of *UR* 8-9, 18, 28-29
   Suggested names for (*see* Georgium Sidus *and* Herschel)
   (*see also the book* **Uranus: The Sideways Planet**)
Ursa Major (constellation) *GUIDE* 9, 11; *MYTH* 17, 21, 23 (*see also* Big Dipper)
US (*see* United States)
USS Enterprise (fictional starship) *SF* 11 (*see also* "Star Trek")
USSR (*see* Soviet Union)
Utah (US) *RPS* 15

V

V-2 (German rocket) *PILOT* 5; *RPS* 8; *WSP* 5, 7, 28
Valles Marineris (Martian canyon) *MARS* 10-11
Variable stars *ANCIENT* 11; *GUIDE* 14; *MYTH* 25; *TODAY* 18, 19; *UNIV* 13 (*see also* Algol, Mira, *and* Pulsars)
Vega (former pole star) *GUIDE* 12-13; *MYTH* 25
Vega (Soviet probe) *COMET* 27; *RPS* 16, 21
Vela, Rudolfo Neri (cosmonaut) *WSP* 29
Venera (Soviet probes) *GARB* 21; *VENUS* 14, 15, 21; *WSP* 21
   Venera 4 *VENUS* 13
   Venera 13 *VENUS* 13
Venus (goddess of love and beauty) *MYTH* 8-9, 18; *VENUS* 5
Venus (planet) *ANCIENT* 8-9, 13, 15, 16, 27; *ASTER* 4, 6-7, 18; *EARTH* 18, 21, 27, 28-29; *GARB* 6, 21, 29; *GUIDE* 18, 20-21, 22; *JUP* 4, 29; *LIFE* 13, 14, 16; *MARS* 6, 29; *MERC* 7, 8, 15, 23, 26, 28-29; *MYTH* 8; *NEP* 7, 28; *PILOT* 22, 27; *PROJ* 12; *RPS* 20-21, 26; *SAT* 28-29; *SF* 14; *SOL* 8, 10, 12-13, 18; *STAR* 17, 22; *SUN* 29; *UFO* 14, 15, 29; *UNIV* 6, 25; *UR* 28; *WSP* 5, 20-21
   Atmosphere of *EARTH* 18; *LIFE* 16; *PILOT* 27; *RPS* 20-21; *SF* 14; *VENUS* 8-9, 12-13, 14-15, 17, 20-21, 24-25, 26-27, 28
   Axis of *VENUS* 19
   Clouds of *EARTH* 18, 27; *GUIDE* 21; *LIFE* 16; *RPS* 20-21; *VENUS* 8-9, 10, 12-13, 15, 25, 28-29
   "Continents" of (*see* Aphrodite Terra *and* Ishtar Terra)
   Core of *MERC* 8; *VENUS* 12-13, 20
   As Earth's twin *LIFE* 13, 16; *VENUS* 10-11, 17, 25, 27, 28
   Magnetic field, lack of *VENUS* 20
   Mountains of *VENUS* 22-23
   Oceans of *VENUS* 10, 21, 25
   Orbit of *MERC* 7; *VENUS* 9, 11, 29
   Phases of *MERC* 15; *VENUS* 6-7, 9, 25
   Reverse rotation of *RPS* 20; *VENUS* 3, 17, 19, 20, 28
   Swamps on *SF* 14; *VENUS* 10-11
   Weather on *RPS* 21; *VENUS* 17
   (*see also the book* **Venus: A Shrouded Mystery**)
"Venusian flying saucer" *UFO* 17
Venusian weather balloon (probe) *RPS* 21
"Venusians" (*see* Alien life)
Verne, Jules (French author) *SF* 4, 8, 12, 28, 29; *WSP* 25 (*see also* Five Weeks in a Balloon, From the Earth to the Moon, In the Twenty-Ninth Century — The Day of an American Journalist, *and* Twenty Thousand Leagues Under the Sea)
Very Large Array (radio telescopes) *TODAY* 13
Vesta (goddess) *ASTER* 11
Vesta (asteroid) *ASTER* 11, 16, 28, 29
Vietnam *WSP* 29
Viking (probes) *GARB* 20; *LIFE* 10, 14-15;

MARS 5, 11, 12, 13, 14-15, 17, 20; PILOT 27; RPS 22-23
    Viking 1 MARS 11, 12, 17; RPS 22-23
    Viking 2 MARS 12; RPS 22
Viking (satellite) WSP 29
Vikings (*see* Norse, ancient)
Violet shift TODAY 28, 29 (*see also* Red shift)
Virginia (US) SUN 20; UFO 28
Virgo (constellation) GAL 16; GUIDE 10-11, 19
Virgo cluster (galactic cluster) GAL 16-17
Volcanoes and volcanic activity COMET 24-25; EARTH 12-13, 14, 18-19; JUP 22-23, 24-25; MARS 10-11, 12, 20, 24-25; MERC 10-11; MOON 19, 28-29; RPS 22, 24; VENUS 10, 16-17, 23
    On Earth COMET 24-25; EARTH 12-13, 14; JUP 22 (*see also* Krakatau, Parícutin, "Ring of Fire," *and* St. Helens, Mt.)
    On Io EARTH 18-19; JUP 22-23, 24-25; RPS 24
    On Mars MARS 10-11, 20, 24-25; RPS 22 (*see also* Olympus Mons *and* Hecates Tholus)
    On Moon MERC 10-11; MOON 19, 28-29
    On Venus VENUS 10, 16-17, 23 (*see also* "Ice volcanoes")
Volcanoes, ice (*see* "Ice volcanoes")
Von Braun, Wernher (German-American scientist) PILOT 5; SF 17; WSP 5
Vostok 1 (rocket) WSP 28
Voyage to the Moon, A (novel) SF 24-25 (*see also* Bergerac, Cyrano de)
Voyager probes GARB 24-26, 28-29; JUP 9, 12-13, 14-15, 16-17, 20-21, 27; LIFE 9; NEP 8-9, 10-11, 12-13, 15, 16-17, 19, 20-21, 25, 27, 28-29; PILOT 25, 26; PLUTO 23; PROJ 28; RPS 16, 24-26; SAT 12-13, 14, 16, 18-19; SOL 19, 21; UR 4, 6-7, 14, 16-17, 18-19, 20, 22, 25, 26, 29; WSP 21
    Voyager 1 GARB 24-25; JUP 9, 12, 14-15, 17; LIFE 9; NEP 16-17, 28-29; PROJ 28; SAT 19; SOL 19, 21; UR 16
    Voyager 2 GARB 24-26; JUP 9, 12-13, 20-21; LIFE 9; NEP 8-9, 10-11, 12-13, 15, 16-17, 19, 20-21, 25, 27, 28-29; PILOT 25, 26; PLUTO 23; PROJ 28; RPS 25, 26; SAT 13, 18; UR 4, 6-7, 14, 16-17, 18-19, 20, 22, 25, 26, 29; WSP 21
Vulcan (Roman god of fire) MERC 24-25
Vulcan (imaginary planet) MERC 24-25

W

Wallops Island, Virginia (tracking station) SUN 20
War in the Air, The (novel) SF 4 (*see also* Wells, H. G.)
War of the Worlds, The (novel by H. G. Wells) LIFE 12-13; UFO 17
"War of the Worlds, The" (radio play by Orson Welles) MARS 7; UFO 17
Warminster, Wiltshire (Britain) UFO 12-13
Washington, George (US president) MYTH 28
Washington, DC (US) MYTH 28; TODAY 24-25
Washington State (US) EARTH 12; MYTH 28
Washingtonia (asteroid) ASTER 11
Water Star (ancient Chinese name for Mercury) MYTH 8 (*see also* Mercury [planet])
Water, water ice, and water vapor COL 8-9; COMET 25; EARTH 6, 8, 9, 10, 13, 15, 19, 20, 22, 27; GARB 7, 16; JUP 8-9, 18, 21; LIFE 5, 10, 13, 14, 17, 28; MARS 6, 7, 12, 21, 22, 24, 25, 26-27; MERC 10; MOON 6, 19, 25; PILOT 7, 11, 20-21, 29; PLUTO 19; PROJ 6, 8, 14; RPS 13, 20, 23; SAT 9-10; SF 14; SOL 14, 16; SUN 12, 24; TODAY 14-15; UFO 5; UR 17; VENUS 10, 25, 27; WSP 24
Weather
    On Earth EARTH 15; NEP 15; WSP 8, 9, 19, 23
    On Jupiter JUP 10-11, 13, 24-25; NEP 15; UR 18
    On Mars MARS 10, 21
    On Neptune NEP 15
    On Saturn NEP 15; UR 18
    On Uranus NEP 15; UR 18
    On Venus RPS 21; VENUS 17
Weather balloons RPS 21; UFO 14, 23
Weather satellites GARB 15, 29; RPS 12-13, 14, 28-29; WSP 8-9 (*see also* Nimbus *and* Tiros 8)
Weightlessness PILOT 14-15, 18-19, 20, 22, 28, 29; SF 10-11; WSP 17 (*see also* G-force, Gravity and gravitational pull, Gravity, artificial, *and* Zero gravity)
Welles, Orson (US actor) MARS 7; UFO 17
Wells, H. G. (British author) LIFE 12-13; SF 4, 6, 12, 28, 29; UFO 17; WSP 25 (*see also First Men in the Moon, The, Time Machine, The, War in the Air, The, War of the Worlds, The* [novel], *and World Set Free, The*)
West Germany SUN 24; WSP 29 (*see also* Germany and the Germans)
Whales LIFE 22
Whirlpool Galaxy GAL 20; GUIDE 26
White, Edward (astronaut) PILOT 6-7

White dwarfs (stars) COL 21; GAL 8-9; GUIDE 25; QPB 10, 11, 14; SOL 24-25, 29; STAR 20-21, 24, 29; UNIV 26
White light PROJ 14; TODAY 28, 29
Will o' the wisp UFO 14
William the Conqueror (king of England) COMET 21, 28
Wisconsin (US) UFO 11
Wolf Creek Site (Australia) DINOS 9
Women in space (*see* McAuliffe, Christa; Resnik, Judith; Ride, Sally; Savitskaya, Svetlana; *and* Tereshkova, Valentina)
Wood Star (ancient Chinese name for Jupiter) MYTH 8 (*see also* Jupiter [planet])
Woomera (Australia) WSP 28, 29
Worcester (Massachusetts) RPS 7
World Set Free, The (novel) SF 4, 28 (*see also* Wells, H. G.)
World War II PILOT 5; RPS 8; UFO 8-9; WSP 5, 7, 28
Wresat (satellite) WSP 29
Wyoming (US) ASTER 19; TODAY 5

X

X-rays DINOS 6, 29; GAL 24; MERC 16; PROJ 14, 18; QPB 18-19, 21; TODAY 28, 29; VENUS 13
X-Wing Fighters (fictional spacecraft) SF 26 (*see also Star Wars* [movie])

Y

Yerkes Observatory (Wisconsin) TODAY 16-17, 23
Yucatán (Mexico) ANCIENT 8, 29; RPS 13

Z

Zero gravity PILOT 18-19, 28-29; SF 10; WSP 13 (*see also* Weightlessness)
Zodiac (constellations) GUIDE 10-11, 18-19; JUP 6; MYTH 18-19, 27 (*see also* Aquarius, Aries, Cancer, Capricorn, Gemini, Leo, Libra, Pisces, Sagittarius, Scorpio, Taurus, *and* Virgo)
Zodiac, Aztec MYTH 19
Zodiac, Chinese MYTH 19
Zodiac, Egyptian (*see* Zodiac of Dendera)
Zodiac, Japanese MYTH 19
Zodiac of Dendera GUIDE 11
Zond-3 (Soviet lunar probe) MOON 15

Isaac Asimov's Library of the Universe

Ancient Astronomy
The Asteroids
Astronomy Today
The Birth and Death of Stars
Colonizing the Planets and Stars
Comets and Meteors
Did Comets Kill the Dinosaurs?
Earth: Our Home Base
The Earth's Moon
How Was the Universe Born?
Is There Life on Other Planets?
Jupiter: The Spotted Giant
Mars: Our Mysterious Neighbor
Mercury: The Quick Planet
Mythology and the Universe
Neptune: The Farthest Giant
Our Milky Way and Other Galaxies
Our Solar System
Piloted Space Flights
Pluto: A Double Planet?
Projects in Astronomy
Quasars, Pulsars, and Black Holes
Rockets, Probes, and Satellites
Saturn: The Ringed Beauty
Science Fiction, Science Fact
Space Garbage
The Space Spotter's Guide
The Sun
Unidentified Flying Objects
Uranus: The Sideways Planet
Venus: A Shrouded Mystery
The World's Space Programs